COUNTING

COUNTING

Koh Khee Meng
National University of Singapore, Singapore

Tay Eng Guan
National Institute of Education
Nanyang Technological University, Singapore

World Scientific
New Jersey • London • Singapore • Hong Kong

Published by

World Scientific Publishing Co. Pte. Ltd.

P O Box 128, Farrer Road, Singapore 912805

USA office: Suite 1B, 1060 Main Street, River Edge, NJ 07661

UK office: 57 Shelton Street, Covent Garden, London WC2H 9HE

British Library Cataloguing-in-Publication Data
A catalogue record for this book is available from the British Library.

The authors and publisher would like to thank the following organizations for their permission to reproduce the selected problems in this book:

The Cambridge International Examinations
The MAA American Mathematics Competitions

COUNTING

ISBN 981-238-063-9
ISBN 981-238-064-7 (pbk)

Printed in Singapore by Mainland Press

Preface

Combinatorics is a branch of mathematics dealing with discretely structured problems. Its scope of study includes selections and arrangements of objects with prescribed conditions, configurations involving a set of nodes interconnected by edges (called graphs), and designs of experimental schemes according to specified rules. Combinatorial problems and their applications can be found not only in various branches of mathematics, but also in other disciplines such as engineering, computer science, operational research, management sciences and the life sciences. Since computers require discrete formulation of problems, combinatorial techniques have become essential and powerful tools for engineers and applied scientists, in particular, in the area of designing and analyzing algorithms for various problems which range from designing the itineraries for a shipping company to sequencing the human genome in the life sciences.

The *counting problem*, which seeks to find out how many arrangements there are in a particular situation, is one of the basic problems in combinatorics. Counting has been used in the social sciences for calculating the Shapley-Shubik power index which measures the power of a player in a decision-making body (such as a shareholders' meeting, a parliament or the UN Security Council). In Chemistry, Cayley used graphs to count the number of isomers of saturated hydrocarbons; while Harary and Read counted the number of certain organic compounds built up from benzene rings by representing them as configurations of hexagons joined together along a common edge. In Genetics, by counting all possibilities for a DNA chain made up of the four bases, scientists arrive at an astoundingly large number and so are able to understand the tremendous possible variation in genetic makeup. Counting has been used as well to study the primary and secondary structures of RNA.

This booklet is intended as an introduction to basic counting techniques for upper secondary and junior college students, and teachers. We believe that it would also be of interest to those who appreciate mathematics and to avid puzzle-solvers.

The variety of problems and applications in this booklet is not only useful for building up an aptitude in counting but is a rich source for honing basic skills and techniques in general problem-solving. Many of the problems evade routine and, as a desired result, force the reader to think hard in his attempts to solve them. In fact, the diligent reader will often discover more than one way of solving a particular problem, which is indeed an important awareness in problem-solving. This booklet thus helps to provide students an early start to learning problem-solving heuristics and thinking skills.

The first two chapters cover two basic principles, namely, the Addition Principle and the Multiplication Principle. Both principles are commonly used in counting, even by those who would never count themselves as students of mathematics! However, these principles have equally likely been misunderstood and misused. These chapters help to avoid this by stating clearly the conditions under which the principles can be applied. Chapter 3 introduces the concepts of combinations and permutations by viewing them as subsets and arrangements of a set of objects, while Chapter 4 provides various applications of the concepts learnt.

Many apparently complex counting problems can be solved with just "a change of perspective". Chapter 5 presents an important principle along this line, i.e. the Bijection Principle; while Chapter 6 introduces a very useful perspective to which many counting problems can be converted to, i.e. the distribution of balls into boxes. The next three chapters flesh out the Bijection Principle and the distribution perspective with a number of applications and variations.

In Chapter 3, we introduce a family of numbers which are denoted by $\binom{n}{r}$ or C_r^n. The last three chapters put this family of numbers in the context of the binomial expansion and Pascal's Triangle. A number of useful identities are proven and problems are posed where these identities surprisingly appear.

Chapter 13 closes this booklet with a collection of interesting problems in which the approaches to solving them appear as applications of one or more concepts learnt in all the earlier sections. Problems in this and other sections marked with (C) are reproduced by permission of the University of Cambridge Local Examinations Syndicate and those with (AIME) are from the American Invitational Mathematics Examination. We would like to express our gratitude to the above organizations for allowing us to include these problems in the book.

This booklet is based on the first six from a series of articles on counting that first appeared in Mathematical Medley, a publication of the Singapore Mathematical Society. We would like to thank Tan Ban Pin who greatly helped the first author with the original series. Many thanks also to our colleagues, Dong Fengming, Lee Tuo Yeong and Toh Tin Lam for reading through the draft and checking through the problems — any mistakes that remain are ours alone.

For those who find this introductory work interesting and would like to know more about the subject, a recommended list of publications for further reading is provided at the end of this book.

Koh Khee Meng
Tay Eng Guan

Contents

Chapter 1

The Addition Principle

In the process of solving a counting problem, there are two very simple but basic principles that we always apply. They are called the *Addition Principle* and the *Multiplication Principle*. In this chapter, we shall introduce the former and illustrate how it is applied.

Let us begin with a simple problem. Consider a 4-element set $A = \{a, b, c, d\}$. In how many ways can we form a 2-element subset of A? This can be answered easily by simply listing all the 2-element subsets:

$$\{a,b\}, \{a,c\}, \{a,d\}, \{b,c\}, \{b,d\}, \{c,d\}.$$

Thus, the answer is 6.

Let us try a slightly more complicated problem.

Example 1.1 *A group of students consists of 4 boys and 3 girls. How many ways are there to select 2 students of the same sex from the group?*

Solution As the problem requires us to select students of the *same sex*, we naturally divide our consideration into two distinct cases: both of the two students are boys, or, both are girls. For the former case, this is the same as selecting 2 elements from a 4-element set. Thus, as shown in the preceding discussion, there are 6 ways. For the latter case, assume the 3 girls are g_1, g_2 and g_3. Then there are 3 ways to form such a pair; namely,

$$\{g_1, g_2\}, \{g_1, g_3\}, \{g_2, g_3\}.$$

Thus, the desired number of ways is $(6 + 3)$, which is 9. □

In dealing with counting problems that are not so straightforward, we quite often have to divide our consideration into cases which are *disjoint* (like boys or girls in Example 1.1) and *exhaustive* (besides boys or girls, no other cases remain). Then the total number of ways would be the sum of the numbers of ways from each case. More precisely, we have:

The Addition Principle

Suppose that there are n_1 ways for the event E_1 to occur and n_2 ways for the event E_2 to occur. If all these ways are distinct, then the number of ways for E_1 or E_2 to occur is $n_1 + n_2$.

(1.1)

For a finite set A, the *size* of A or the *cardinality* of A, denoted by $|A|$, is the number of elements in A. For instance, if $A = \{u, v, w, x, y, z\}$, then $|A| = 6$; if A is the set of all the letters in the English alphabet, then $|A| = 26$; if ϕ denotes the *empty* (or *null*) set, then $|\phi| = 0$.

Using the language of sets, the Addition Principle simply states the following.

If A and B are finite sets with $A \cap B = \phi$, then $|A \cup B| = |A| + |B|$.

(1.2)

Two sets A and B are *disjoint* if $A \cap B = \phi$. Clearly, the above result can be extended in a natural way to any finite number of pairwise disjoint finite sets as given below.

(AP) If $A_1, A_2, \ldots, A_n, n \geq 2$, are finite sets which are pairwise disjoint, i.e. $A_i \cap A_j = \phi$ for all i, j with $1 \leq i < j \leq n$, then

$$|A_1 \cup A_2 \cup \cdots \cup A_n| = |A_1| + |A_2| + \cdots + |A_n|,$$

or, in a more concise form:

$$\left| \bigcup_{i=1}^{n} A_i \right| = \sum_{i=1}^{n} |A_i|.$$

(1.3)

Example 1.2 *From town X to town Y, one can travel by air, land or sea. There are 3 different ways by air, 4 different ways by land and 2 different ways by sea as shown in Figure 1.1.*

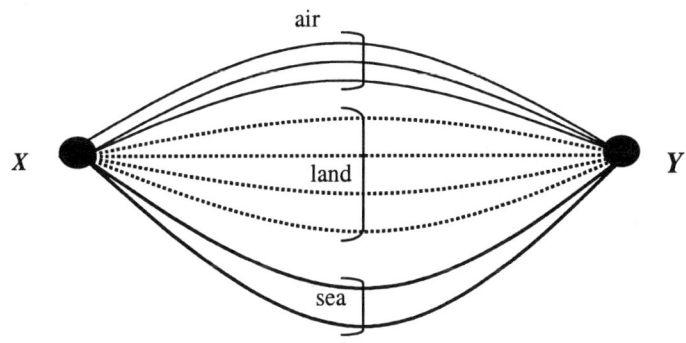

Figure 1.1

How many ways are there from X to Y?

Let A_1 be the set of ways by air, A_2 the set of ways by land and A_3 the set of ways by sea from X to Y. We are given that

$$|A_1| = 3, \quad |A_2| = 4 \quad \text{and} \quad |A_3| = 2.$$

Note that $A_1 \cap A_2 = A_1 \cap A_3 = A_2 \cap A_3 = \phi$ and $A_1 \cup A_2 \cup A_3$ is the set of ways from X to Y. Thus, the required number of ways is $|A_1 \cup A_2 \cup A_3|$, which, by (AP), is equal to

$$|A_1| + |A_1| + |A_1| = 3 + 4 + 2 = 9. \qquad \square$$

Example 1.3 *Find the number of squares contained in the 4×4 array (where each cell is a square) of Figure 1.2.*

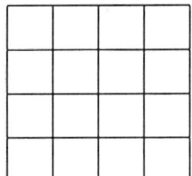

Figure 1.2

Solution   The squares in the array can be divided into the following 4 sets:

A_1: the set of 1×1 squares,
A_2: the set of 2×2 squares,
A_3: the set of 3×3 squares, and
A_4: the set of 4×4 squares.

There are sixteen "1×1 squares". Thus $|A_1| = 16$. There are nine "2×2 squares". Thus $|A_2| = 9$. Likewise, $|A_3| = 4$ and $|A_4| = 1$.

Clearly, the sets A_1, A_2, A_3, A_4 are pairwise disjoint and $A_1 \cup A_2 \cup A_3 \cup A_4$ is the set of the squares contained in the array of Figure 1.2. Thus, by (AP), the desired number of squares is given by

$$\left| \bigcup_{i=1}^{4} A_i \right| = \sum_{i=1}^{4} |A_i| = 16 + 9 + 4 + 1 = 30. \qquad \square$$

Example 1.4   *Find the number of routes from X to Y in the one-way system shown in Figure 1.3.*

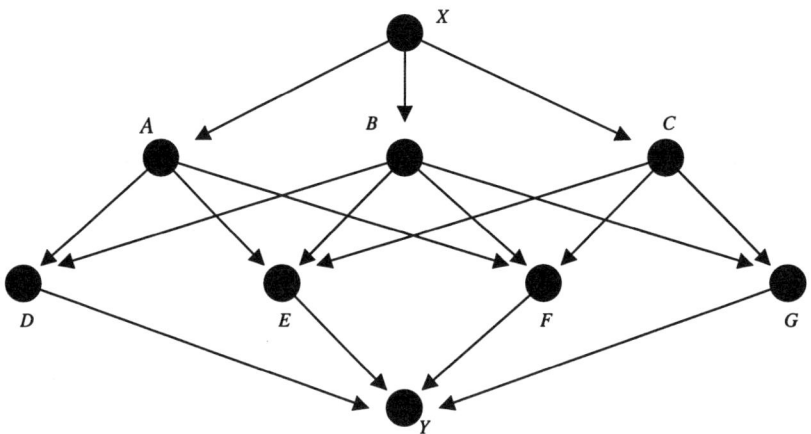

Figure 1.3

Solution   Of course, one can count the number of such routes by simply listing all of them: $X \to A \to D \to Y$, $X \to A \to E \to Y$, ..., $X \to C \to G \to Y$.

Let us, however, see how to apply (AP) to introduce a more general method.

Call a route from X to Y an X—Y *route*. It is obvious that just before reaching Y along any X—Y route, one has to reach D, E, F or G. Thus, by (AP), the number of X—Y routes is the sum of the numbers of X—D routes, X—E routes, X—F routes and X—G routes.

How many X—D routes are there? Just before reaching D along any X—D route, one has to reach either A or B, and thus, by (AP), the number of X—D routes is the sum of the numbers of X—A routes and X—B routes. The same argument applies to others (X—E routes, ...) as well.

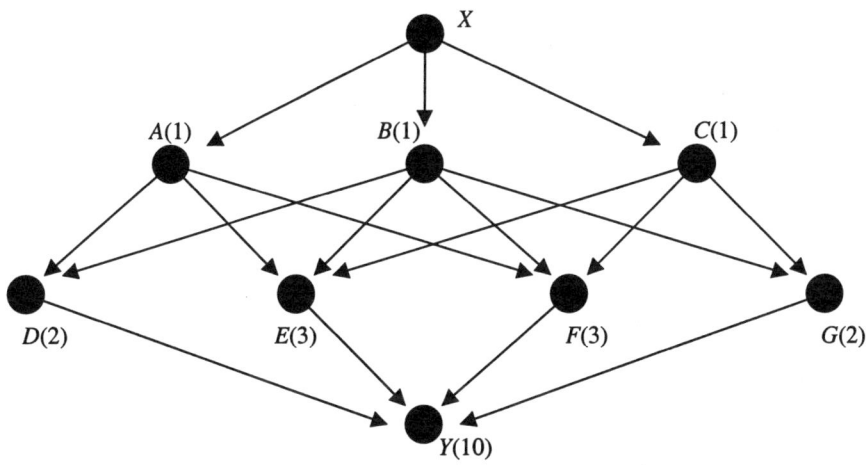

Figure 1.4

It is clear that the number of X—A routes (X—B routes and X—C routes) is 1. With these initial values, one can compute the numbers of X—D routes, X—E routes, etc., using (AP) as explained above. These are shown in brackets at the respective vertices in Figure 1.4. Thus, we see that the total number of possible X—Y routes is $2 + 3 + 3 + 2$, *i.e.* 10. □

Exercise

1.1 We can use 6 pieces of  to cover a 6 × 3 rectangle, for example, as shown below:

In how many different ways can the 6 × 3 rectangle be so covered?

1.2 Do the same problem as in Example 1.3 for 1 × 1, 2 × 2, 3 × 3 and 5 × 5 square arrays. Observe the pattern of your results. Find, in general, the number of squares contained in an $n \times n$ square array, where $n \geq 2$.

1.3 How many squares are there in

   (i) the following 4 × 3 array (where each cell is a square)?

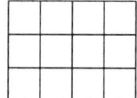

   (ii) an $n \times 3$ array (where each cell is a square), with $n \geq 5$?

1.4 How many squares are there in the following array (where each cell is a square)?

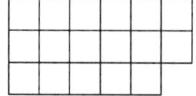

1.5 Find the number of triangles in the following figure.

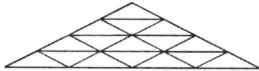

1.6 Find the number of triangles in the following figure.

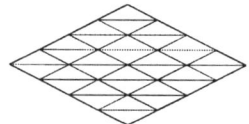

1.7   How many squares are there in the following configuration (where each cell is a square with diagonals)?

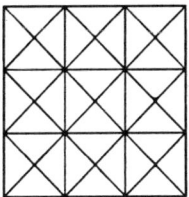

1.8   Following the arrows given in the diagram, how many different routes are there from N to S?

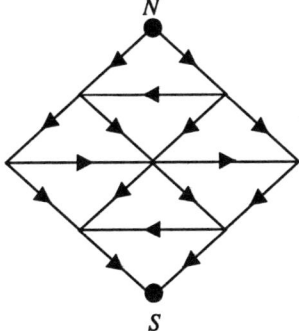

1.9   Following the arrows given in the diagram, how many different routes are there from N to S?

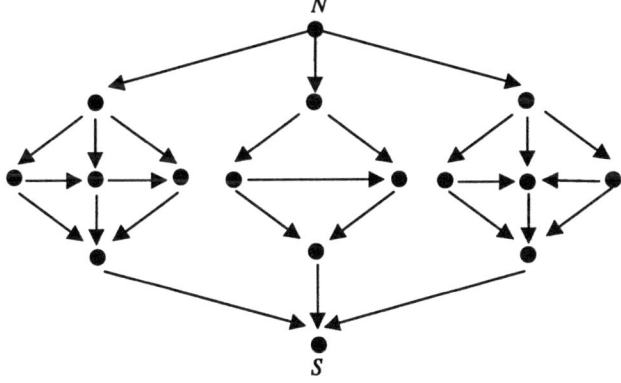

Chapter 2

The Multiplication Principle

Mr. Tan is now in town X and ready to leave for town Z by car. But before he can reach town Z, he has to pass through town Y. There are 4 roads (labeled 1, 2, 3, 4) linking X and Y, and 3 roads (labeled as a, b, c) linking Y and Z as shown in Figure 2.1. How many ways are there for him to drive from X to Z?

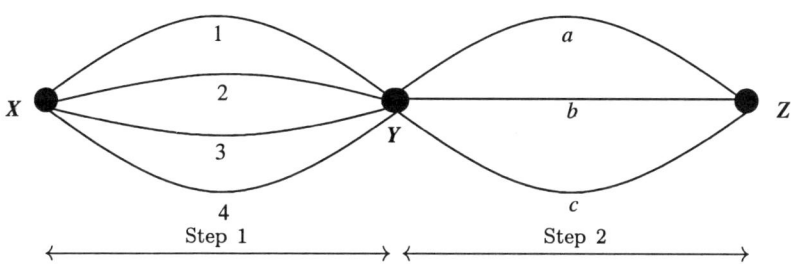

Figure 2.1

Mr. Tan may choose road "1" to leave X for Y, and then select "a" from Y to Z. For simplicity, we denote such a sequence by $(1, a)$. Thus, by listing all possible sequences as shown below:

$$(1, a), (1, b), (1, c),$$
$$(2, a), (2, b), (2, c),$$
$$(3, a), (3, b), (3, c),$$
$$(4, a), (4, b), (4, c),$$

we get the answer $(4 \times 3 =)12$.

Very often, to accomplish a task, one may have to split it into ordered stages and then complete the stages step by step. In the above example, to leave X and reach Z, Mr. Tan has to split his journey into 2 stages: first from X to Y, and then Y to Z. There are 4 roads to choose from in Step 1: To each of these 4 choices, there are 3 choices in Step 2. Note that the number of choices in Step 2 is independent of the number of choices in Step 1. Thus, the number of ways from X to Z is given by $4 \times 3 \ (= 12)$. This illustrates the meaning of the following principle.

The Multiplication Principle

Suppose that an event E can be split into two events E_1 and E_2 in ordered stages. If there are n_1 ways for E_1 to occur and n_2 ways for E_2 to occur, then the number of ways for the event E to occur is $n_1 n_2$.   (2.1)

Example 2.1 *How many ways are there to select 2 students of different sex from a group of 4 boys and 3 girls?*

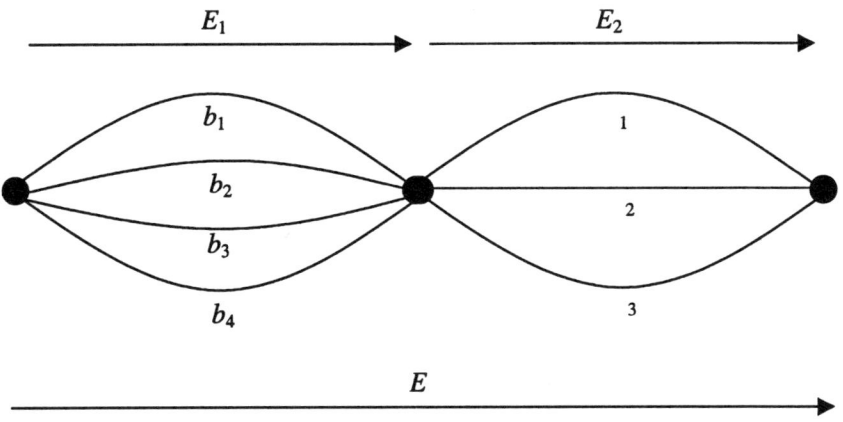

E : forming a pair consisting of a boy and a girl;
E_1: selecting a boy;
E_2: selecting a girl.

Figure 2.2

Solution   The situation when the 2 students chosen are of the same sex was discussed in Example 1.1. We now consider the case where the 2 students chosen are of different sex. To choose 2 such students, we may first choose a boy and then select a girl. There are 4 ways for step 1 and 3 ways for step 2 (see Figure 2.2). Thus, by the Multiplication Principle, the answer is 4×3 ($= 12$). $\square$

The Addition Principle can be expressed using set language. The Multiplication Principle can likewise be so expressed. For the former, we make use of the *union* $A \cup B$ of sets A and B. For the latter, we shall introduce the *cartesian product* $A \times B$ of sets A and B. Thus given two sets A and B, let

$$A \times B = \{(x, y) : x \in A,\ y \in B\};$$

namely, $A \times B$ consists of all *ordered pairs* (x, y), where the first coordinate, "x", is any member in the first set A, and the second coordinate, "y", is any member in the second set B. For instance, if $A = \{1, 2, 3, 4\}$ and $B = \{a, b, c\}$, then

$$A \times B = \{(1, a), (1, b), (1, c), (2, a), (2, b), (2, c), (3, a), (3, b),$$
$$(3, c), (4, a), (4, b), (4, c)\}.$$

Assume that A and B are finite sets. How many members (i.e. ordered pairs) are there in the set $A \times B$? In forming ordered pairs in $A \times B$, a member, say "x" in A is paired up with every member in B. Thus there are $|B|$ ordered pairs having "x" as the first coordinate. Since there are $|A|$ members in A, altogether we have $|A|\,|B|$ ordered pairs in $A \times B$. That is,

$$\boxed{|A \times B| = |A|\,|B|.} \qquad (2.2)$$

Principle (2.1) and result (2.2) are two different forms of the same fact. Indeed, an event E which is split into two events in ordered stages can be regarded as an ordered pair (a, b), where "a" stands for the first event and "b" the second; and vice versa.

Likewise, Principle (2.1) can be extended in a very natural way as follows:

(MP)   Suppose that an event E can be split into k events $E_1, E_2, \ldots, E_k$ in ordered stages. If there are n_1 ways for E_1 to occur, n_2 ways for E_2 to occur, $\ldots$, and n_k ways for E_k to occur, then the number of ways for the event E to occur is given by $n_1 n_2 \ldots n_k$.   (2.3)

By extending the cartesian product $A \times B$ of two sets to $A_1 \times A_2 \times \cdots \times A_k$ of k sets, we shall also derive an identity which extends (2.2) and expresses (2.3) using set language.

Let $A_1, A_2, \ldots, A_k$ be k finite sets, and let

$$A_1 \times A_2 \times \cdots \times A_k$$
$$= \{(x_1, x_2, \ldots, x_k) : x_i \in A_i \quad \text{for each } i = 1, 2, \ldots, k\}.$$

Then

(MP) $|A_1 \times A_2 \times \cdots \times A_k| = |A_1||A_2|\ldots|A_k|.$ (2.4)

Example 2.2   *There are four 2-digit binary sequences:* $00, 01, 10, 11.$ *There are eight 3-digit binary sequences:* $000, 001, 010, 011, 100, 101, 110, 111.$ *How many 6-digit binary sequences can we form?*

Solution   The event of forming a 6-digit binary sequence can be split into 6 ordered stages as shown in Figure 2.3.

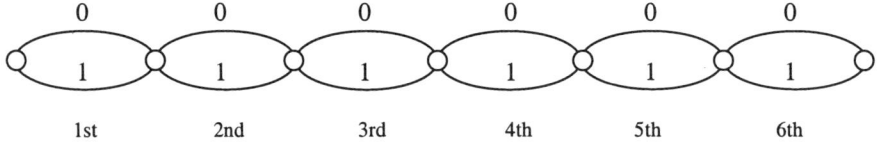

Figure 2.3

Thus, by (2.3), the desired number of sequences is $2 \times 2 \times 2 \times 2 \times 2 \times 2 = 2^6$.

Using set language, the same problem can be treated as follows. We have

$$A_1 = A_2 = \cdots = A_6 = \{0, 1\}.$$

The members in $A_1 \times A_2 \times \cdots \times A_6$ can be identified with 6-digit binary sequences in the following way:

$$(1,1,0,1,0,1) \leftrightarrow 110101,$$

$$(0,1,1,0,1,1) \leftrightarrow 011011,$$

$$\text{etc.}$$

Thus, the number of 6-digit binary sequences is given by $|A_1 \times A_2 \times \cdots \times A_6|$, which, by (2.4), is equal to

$$|A_1||A_2|\ldots|A_6| = 2 \times 2 \times 2 \times 2 \times 2 \times 2 = 2^6 \,. \qquad \square$$

From now on, (MP) shall refer to Principle (2.3) or the identity (2.4).

Example 2.3 *Figure 2.4 shows 9 fixed points $a, b, c, \ldots, i$ which are located on the sides of $\triangle ABC$. If we select one such point from each side and join the selected points to form a triangle, how many such triangles can be formed?*

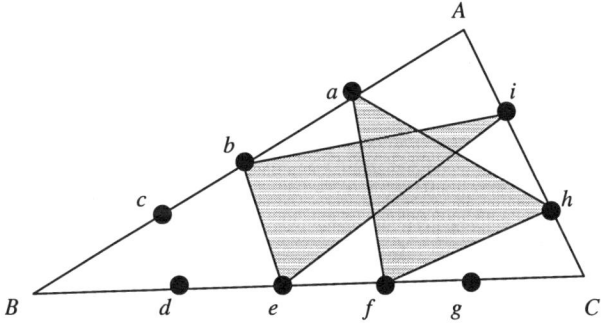

Figure 2.4

Solution To form such a triangle, we first select a point on AB, then a point on BC and finally a point on CA. There are 3 ways in step 1 (one of a, b, c), 4 ways in step 2 (one of d, e, f, g) and 2 ways in step 3 (either h or i). Thus by (MP), there are $3 \times 4 \times 2$ ($= 24$) such triangles. $\qquad \square$

We have seen in both the preceding and the current chapters some problems that can be solved by applying (AP) or (MP) *individually*. Indeed, more often than not, problems that we encounter are more complex and these require that we apply the principles together. The following is an example.

Example 2.4 (Continuation of Example 2.3) *Find the number of triangles that can be formed using the 9 fixed points of Figure 2.4 as vertices.*

Solution   This problem is clearly more complex than that of Example 2.3 as there are other triangles whose three vertices are not necessarily chosen from three different sides; but then, where else can they be chosen from? The answer is: two from one side and one from the remaining two sides. In view of this, we shall now classify the required triangles into the following two types.

Type 1 — Triangles whose three vertices are chosen from three different sides.

As shown in Example 2.3, there are $3 \times 4 \times 2$ ($= 24$) such triangles.

Type 2 — Triangles having two vertices from one side and one from the other two sides.

We further split our consideration into three subcases.

  (i) *Two vertices from AB and one from BC or CA.*

There are 3 ways to choose two from AB (namely, $\{a,b\}, \{a,c\}$ or $\{b,c\}$) and 6 ways to choose one from the other sides (namely, d, e, f, g, h, i). Thus, by (MP), there are 3×6 ($= 18$) such triangles.

 (ii) *Two vertices from BC and one from CA or AB.*

There are 6 ways to choose two from BC (why?) and 5 ways to choose one from the other sides (why?). Thus, by (MP), there are 6×5 ($= 30$) such triangles.

(iii) *Two vertices from CA and one from AB or BC.*

There is only one way to choose two from CA and there are 7 ways to choose one from the other sides. Thus, by (MP), there are 1×7 ($= 7$) such triangles.

Summing up the above discussion, we conclude that by (AP), there are $18 + 30 + 7$ ($= 55$) triangles of Type 2.

As there are 24 triangles of Type 1 and 55 triangles of Type 2, the required number of triangles is thus, by (AP), $24 + 55$ (= 79). □

Exercise

2.1 Following the arrows given in the diagram, how many different routes are there from W to E?

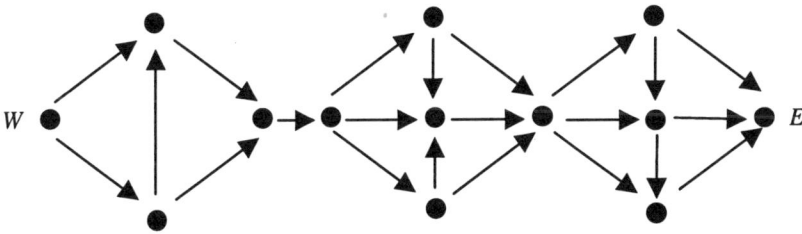

2.2 In the following figure, $ABCD$ and FEC are two perpendicular lines.

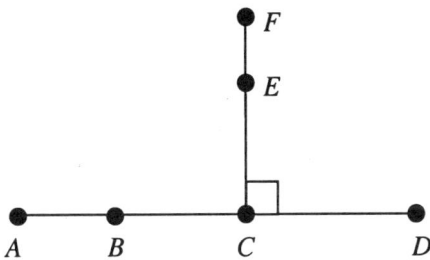

(i) Find the number of right-angled triangles $\triangle XCY$ that can be formed with X, Y taken from A, B, D, E, F.

(ii) Find the number of triangles that can be formed with any three points A, B, C, D, E, F as vertices.

2.3 There are 2 distinct terms in the expansion of $a(p + q)$:

$$a(p + q) = ap + aq.$$

There are 4 distinct terms in the expansion of $(a + b)(p + q)$:

$$(a + b)(p + q) = ap + aq + bp + bq.$$

How many distinct terms are there in each of the expansions of

(i) $(a + b + c + d)(p + q + r + s + t)$,

(ii) $(x_1 + x_2 + \cdots + x_m)(y_1 + y_2 + \cdots + y_n)$, and

(iii) $(x_1 + x_2 + \cdots + x_m)(y_1 + y_2 + \cdots + y_n)(z_1 + z_2 + \cdots + z_t)$?

2.4 In how many different ways can the following configuration be covered by nine 2×1 rectangles?

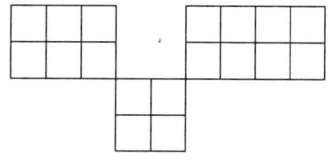

2.5 A ternary sequence is a sequence formed by 0, 1 and 2. Let n be a positive integer. Find the number of n-digit ternary sequences

(i) with no restrictions;

(ii) which contain no "0";

(iii) which contain at most one "0";

(iv) which contain at most one "0" and at most one "1".

2.6 The

following diagram shows 12 distinct points: $a_1, a_2, a_3, b_1, \ldots, b_4$, $c_1, \ldots, c_5$ chosen from the sides of $\triangle ABC$.

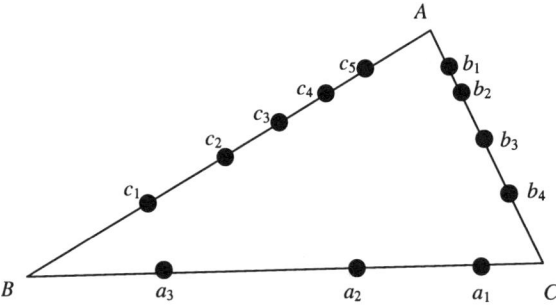

(i) How many line segments are there joining any two points, each point being from a different side of the triangle?

(ii) How many triangles can be formed from these points?

(iii) How many quadrilaterals can be formed from these points?

Chapter 3

Subsets and Arrangements

There are 25 students in the class. How many ways are there to choose 5 of them to form a committee? If among the chosen five, one is to be the chairperson, one the secretary and one the treasurer, in how many ways can this be arranged? In this chapter, our attention will be focused on the counting problems of the above types. We shall see how (MP) is used to solve such problems, and how (MP), by incorporating (AP), enables us to solve more complicated problems.

From now on, for each natural number n, we shall denote by N_n the set of natural numbers from 1 to n inclusive, i.e.,

$$N_n = \{1, 2, 3, \ldots, n\}.$$

Consider the 4-element set $N_4 = \{1, 2, 3, 4\}$. How many subsets of N_4 are there? This question can be answered readily by listing all the subsets of N_4. Table 3.1 lists all the subsets according to the number

Table 3.1

Number of elements	Subsets of N_4	Number of subsets of N_4
0	ϕ	1
1	$\{1\}$, $\{2\}$, $\{3\}$, $\{4\}$	4
2	$\{1, 2\}$, $\{1, 3\}$, $\{1, 4\}$, $\{2, 3\}$, $\{2, 4\}$, $\{3, 4\}$	6
3	$\{1, 2, 3\}$, $\{1, 2, 4\}$, $\{1, 3, 4\}$, $\{2, 3, 4\}$	4
4	$\{1, 2, 3, 4\}$	1

of elements they possess: It is now easy to count the total number of subsets of $N_4(= 16)$.

We note that the 5 numbers, namely, 1, 4, 6, 4, 1 (whose sum is 16) shown in the right hand column of Table 3.1 are the corresponding numbers of r-element subsets of N_4, where $r = 0, 1, 2, 3, 4$. These numbers are very interesting, useful and important in mathematics, and mathematicians have introduced special symbols to denote them.

In general, given two integers n and r with $0 \leq r \leq n$, we denote by $\binom{n}{r}$, the number of r-element subsets of N_n. Thus, Table 3.1 tells us that

$$\binom{4}{0} = 1, \quad \binom{4}{1} = 4, \quad \binom{4}{2} = 6, \quad \binom{4}{3} = 4, \quad \binom{4}{4} = 1.$$

The symbol $\binom{n}{r}$ is read "n choose r". Some other symbols for this quantity include C_r^n and ^nC_r.

Now, what is the value of $\binom{5}{2}$? Since $\binom{5}{2}$ counts, by definition, the number of 2-element subsets of N_5, we may list all these subsets as shown below:

$$\{1,2\}, \{1,3\}, \{1,4\}, \{1,5\}, \{2,3\}, \{2,4\}, \{2,5\}, \{3,4\}, \{3,5\}, \{4,5\},$$

and see that there are 10 in total. Thus, we have $\binom{5}{2} = 10$.

You may ask: How about $\binom{100}{6}$? We are sure that we are too busy to have time to compute $\binom{100}{6}$ by listing all the 6-element subsets of N_{100}. Thus, a natural question arises: Is there a more efficient way to compute $\binom{100}{6}$? The answer is "Yes", and we are going to show you.

Let us first consider a different but related problem. How many ways are there to arrange any three elements of $N_4 = \{1, 2, 3, 4\}$ in a row? With a little patience, we can list all the required arrangements as shown in Table 3.2.

Table 3.2

123	132	213	231	312	321
124	142	214	241	412	421
134	143	314	341	413	431
234	243	324	342	423	432

Thus, there are 24 ways to do so. The answer is "correct" but the method is "naive". Is there a cleverer way to get the answer?

Imagine that we wish to choose 3 numbers from N_4 and put them one by one into 3 spaces as shown.

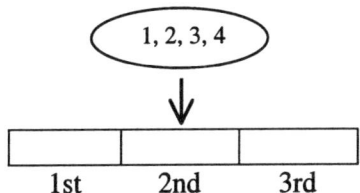

This event can be thought of as a sequence of events: We first select a number from N_4 and place it in the 1st space; we then select another number and place it in the 2nd space; finally, we select another number and place it in the 3rd space. There are 4 choices for the first step, 3 choices (why?) for the second and 2 choices (why?) for the last. Thus, by (MP), there are $4 \cdot 3 \cdot 2 \, (= 24)$ ways to do so. The answer agrees with what we obtained above. Isn't this method better?

This method is better not only in shortening our solution, but also in giving us an idea on how to generalize the above result.

In the above example, we considered the number of ways of arranging 3 elements of N_4 in a row. We now ask a general question: Given integers r, n with $0 \le r \le n$, how many ways are there to arrange any r elements of N_n in a row?

Consider the r spaces shown in Figure 3.1.

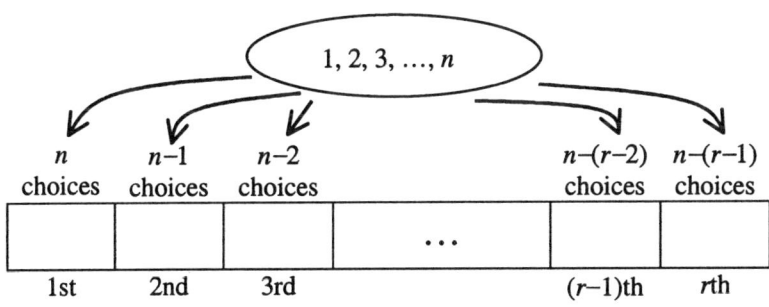

Figure 3.1

We wish to choose r elements from $\{1, 2, \ldots, n\}$ to fill the r spaces, where the ordering of elements matters. There are n choices for the 1st space. After fixing one in the 1st space, there are $n - 1$ choices remaining for the 2nd space. After fixing one in the 2nd space, there are $n - 2$ choices left for the 3rd space, and so on. After fixing one in the $(r - 1)$th space, there are $n - (r - 1)$ choices left for the rth space. Thus, by (MP), the number of ways to arrange any r elements from N_n in a row is given by

$$n(n - 1)(n - 2) \ldots (n - r + 1).$$

For convenience, let us call an arrangement of any r elements from N_n in a row, an *r-permutation* of N_n, and denote by P_r^n the number of r-permutations of N_n. Thus, we have

$$\boxed{P_r^n = n(n - 1)(n - 2) \ldots (n - r + 1).} \tag{3.1}$$

For instance, all the arrangements in Table 3.2 are 3-permutations of N_4, and, by (3.1), the number of 3-permutations of N_4 is given by

$$P_3^4 = 4 \cdot 3 \cdot 2 = 24,$$

which agrees with what we have counted in Table 3.2.

The expression (3.1) for P_r^n looks a bit long. We shall make it more concise by introducing the following useful notation. Given a positive integer n, define $n!$ to be the product of the n consecutive integers $n, n - 1, \ldots, 3, 2, 1$; that is,

$$\boxed{n! = n(n - 1)(n - 2) \ldots 3 \cdot 2 \cdot 1.} \tag{3.2}$$

Thus $4! = 4 \cdot 3 \cdot 2 \cdot 1 = 24$. The symbol "$n!$" is read "$n$ factorial". By convention, we define $0! = 1$.

Using the "factorial" notation, we now have

$$P_r^n = n(n - 1) \ldots (n - r + 1)$$
$$= \frac{n(n - 1) \ldots (n - r + 1)(n - r)(n - r - 1) \ldots 2 \cdot 1}{(n - r)(n - r - 1) \ldots 2 \cdot 1} = \frac{n!}{(n - r)!}.$$

That is,

$$P_r^n = \frac{n!}{(n-r)!} \, . \tag{3.3}$$

When $n = 4$ and $r = 3$, we obtain

$$P_3^4 = \frac{4!}{(4-3)!} = \frac{4!}{1!} = \frac{4 \cdot 3 \cdot 2 \cdot 1}{1} = 4 \cdot 3 \cdot 2 = 24 \, ,$$

which agrees with what we found before.

The expression (3.3) is valid when $0 \le r \le n$. Consider two extreme cases: when $r = 0$ and $r = n$ respectively. When $r = 0$, by (3.3),

$$P_0^n = \frac{n!}{(n-0)!} = \frac{n!}{n!} = 1 \, .$$

(How can this be explained?) When $r = n$, an n-permutation of N_n is simply called a *permutation* of N_n. Thus, by (3.3) and that $0! = 1$, the number of permutations of N_n is given by

$$P_n^n = \frac{n!}{(n-n)!} = \frac{n!}{0!} = n! \, ,$$

i.e.,

$$P_n^n = n! \, . \tag{3.4}$$

Thus, for example, P_5^5 counts the number of permutations of N_5, and we have, by (3.4), $P_5^5 = 5 \cdot 4 \cdot 3 \cdot 2 \cdot 1 = 120$.

Let us now return to the problem of evaluating the quantity $\binom{n}{r}$.

We know from (3.3) that the number P_r^n of r-permutations of N_n is given by $\frac{n!}{(n-r)!}$. We shall now count this number (namely the number of r-permutations of N_n) in a different way.

To get an r-permutation of N_n, we may proceed in the following manner: first select an r-element subset of N_n, and then arrange the chosen r elements in a row. The number of ways for the first step is, by definition, $\binom{n}{r}$, while that for the second step is, by (3.4), $r!$. Thus, by (MP), we have

$$P_r^n = \binom{n}{r} \cdot r! \,.$$

As

$$P_r^n = \frac{n!}{(n-r)!} \,,$$

we have

$$\binom{n}{r} \cdot r! = \frac{n!}{(n-r)!} \,,$$

and thus

$$\binom{n}{r} = \frac{n!}{r!(n-r)!} \,. \qquad (3.5)$$

For instance,

$$\binom{5}{2} = \frac{5!}{2!(5-2)!} = \frac{5!}{2!3!} = 10 \,, \quad \text{while} \quad \binom{100}{6} = \frac{100!}{6!94!} = 1192052400 \,.$$

Note that when $r = 0$ or n, we have

$$\binom{n}{0} = 1 \quad \text{and} \quad \binom{n}{n} = 1 \,.$$

Again, by convention, we define

$$\binom{0}{0} = 1 \,.$$

By applying (3.5), one can show that the following identity holds (see Problem 3.1):

$$\binom{n}{r} = \binom{n}{n-r} \,. \qquad (3.6)$$

Thus, $\binom{10}{8} = \binom{10}{2} = 45$ and $\binom{100}{94} = \binom{100}{6} = 1192052400$.

We define P_r^n as the number of r-permutations and $\binom{n}{r}$ as the number of r-element subsets of N_n. Actually, in these definitions, N_n can be replaced by any n-element set since it is the number of the elements but not the nature of the elements in the set that matters. That is, given any n-element set S, P_r^n (respectively, $\binom{n}{r}$) is defined as the number of r-permutations (respectively, r-element subsets) of S. Any r-element subset of S is also called an r-*combination* of S.

We have introduced the notions of r-permutations (or permutations) and r-combinations (or combinations) of a set S. Always remember that these two notions are closely related but different. While a "combination" of S is just a *subset* of S (and thus the ordering of elements is immaterial), a "permutation" of S is an *arrangement* of certain elements of S (and thus the ordering of elements is important).

Exercise

3.1   Show that for nonnegative integers r and n, with $r \leq n$,

     (i) $\binom{n}{r} = \binom{n}{n-r}$;

     (ii) $r\binom{n}{r} = n\binom{n-1}{r-1}$, where $r \geq 1$;

     (iii) $(n-r)\binom{n}{r} = n\binom{n-1}{r}$;

     (iv) $r\binom{n}{r} = (n-r+1)\binom{n}{r-1}$, where $r \geq 1$.

3.2   Show that for $1 \leq r \leq n$,

     (i) $P_r^{n+1} = P_r^n + rP_{r-1}^n$;

     (ii) $P_r^{n+1} = r! + r(P_{r-1}^n + P_{r-1}^{n-1} + \cdots + P_{r-1}^r)$;

     (iii) $(n-r)P_r^n = nP_r^{n-1}$;

     (iv) $P_r^n = (n-r+1)P_{r-1}^n$;

     (v) $P_r^n = nP_{r-1}^{n-1}$.

3.3   Prove that the product of any n consecutive integers is divisible by $n!$.

3.4   Find the sum

$$1 \cdot 1! + 2 \cdot 2! + 3 \cdot 3! + \cdots + n \cdot n!.$$

Chapter 4

Applications

Having introduced the concepts of r-permutations and r-combinations of an n-element set, and having derived the formulae for P_r^n and $\binom{n}{r}$, we shall now give some examples to illustrate how these can be applied.

Example 4.1 *There are 6 boys and 5 men waiting for their turn in a barber shop. Two particular boys are A and B, and one particular man is Z. There is a row of 11 seats for the customers. Find the number of ways of arranging them in each of the following cases:*

(i) *there are no restrictions;*
(ii) *A and B are adjacent;*
(iii) *Z is at the centre, A at his left and B at his right (need not be adjacent);*
(iv) *boys and men alternate.*

Solution (i) This is the number of permutations of the 11 persons. The answer is 11!.

(ii) Treat $\{A, B\}$ as a single entity. The number of ways to arrange the remaining 9 persons together with this entity is $(9 + 1)!$. But A and B can permute themselves in 2 ways. Thus the total desired number of ways is, by (MP), $2 \cdot 10!$.

(iii)

As shown in the diagram above, A has 5 choices and B has 5 choices as well. The remaining 8 persons can be placed in 8! ways. By (MP), the total desired number of ways is $5 \cdot 5 \cdot 8!$.

(iv) The boys (indicated by b) and the men (indicated by m) must be arranged as shown below.

b	m	b	m	b	m	b	m	b	m	b

The boys can be placed in 6! ways and the men can be placed in 5! ways. By (MP), the desired number of ways is $6!5!$.   □

Example 4.2 *In each of the following cases, find the number of integers between 3000 and 6000 in which no digit is repeated:*

(i) *there are no additional restrictions;*
(ii) *the integers are even.*

Solution   Let $abcd$ be a required integer.

(i) As shown in the diagram below, a has 3 choices (i.e. 3, 4, or 5), say $a = 3$.

$$\{3, 4, 5\}$$

a	b	c	d

Since no digit is repeated, a way of forming "*bcd*" corresponds to a 3-permutation from the 9-element set $\{0, 1, 2, 4, 5, \ldots, 9\}$. Thus the required number of integers is $3 \cdot P_3^9$.

(ii) Again, $a = 3$, 4 or 5. We divide the problem into two cases.

Case (1) $a = 4$ (even)

4	b	c	d

In this case, d has 4 choices (i.e. 0, 2, 6, 8), say $d = 2$. Then a way of forming "*bc*" is a 2-permutation from the 8-element set $\{0, 1, 3, 5, 6, 7, 8, 9\}$. Thus the required number of integers is $4 \cdot P_2^8$.

Case (2) $a = 3$ or 5 (odd)

In this case, d has 5 choices, and the number of ways to form "bc" is P_2^8. The required number of integers is $2 \cdot 5 \cdot P_2^8$.
By (AP), the desired number of integers is

$$4 \cdot P_2^8 + 2 \cdot 5 \cdot P_2^8 = 14 \cdot P_2^8 . \qquad \square$$

Example 4.3 *There are 10 pupils in a class.*

(i) *How many ways are there to form a 5-member committee for the class?*

(ii) *How many ways are there to form a 5-member committee in which one is the Chairperson, one is the Vice-Chairperson, one is the Secretary and one is the Treasurer?*

(iii) *How many ways are there to form a 5-member committee in which one is the Chairperson, one is the Secretary and one is the Treasurer?*

Solution   (i) This is the same as finding the number of 5-combinations of a 10-element set. Thus the answer is $\binom{10}{5} = 252$.
(ii) This is the same as choosing 5 pupils from the class and then placing them in the following spaces.

Chairperson	V–Chairperson	Secretary	Treasurer	Member

Clearly, this is a "permutation" problem, and the answer is $P_5^{10} = \frac{10!}{5!} = 7620480$.
(iii) This problem can be counted in the following procedure: we first select one for Chairperson, then one for Secretary, then one for Treasurer, and finally two from the remainder for committee members as shown below:

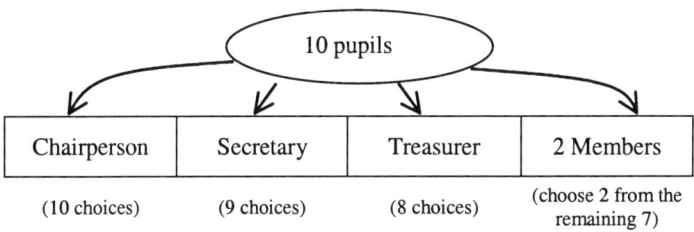

Figure 4.1

Thus, by (MP), the answer is given by $10 \cdot 9 \cdot 8 \cdot \binom{7}{2} = 1814400$.   □

Note There are different ways to solve (iii). You may want to try your own ways.

Example 4.4 *As shown in Example 2.2, the number of 6-digit binary sequences is* 2^6. *How many of them contain exactly two 0's* (*e.g.* $001111, 101101, \ldots$)?

Solution Forming a 6-digit binary sequence with two 0's is the same as choosing two spaces from the following 6 spaces into which the two 0's are put (the rest are then occupied by 1's) as shown below:

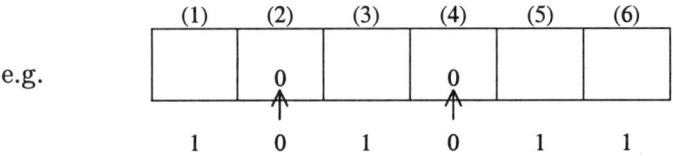

Thus, the number of such binary sequences is $\binom{6}{2}$.   □

Example 4.5 *Figure* 4.2 *shows* 9 *distinct points on the circumference of a circle.*

(i) *How many chords of the circle formed by these points are there?*
(ii) *If no three chords are concurrent in the interior of the circle, how many points of intersection of these chords within the circle are there?*

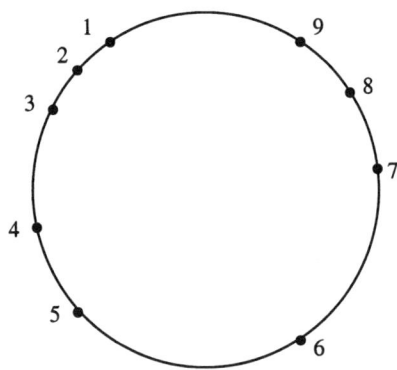

Figure 4.2

Solution (i) Every chord joins two of the nine points, and every two of the nine points determine a unique chord. Thus, the required number of chords is $\binom{9}{2}$.

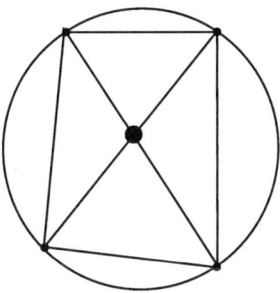

(ii) Every point of intersection of two chords corresponds to four of the nine points, and every four of the nine points determine a point of intersection. Thus the required number of points of intersection is $\binom{9}{4}$. □

Example 4.6 *At a National Wages Council conference, there are 19 participants from the government, the unions and the employers. Among them, 9 are from the unions.*

In how many ways can a 7-member committee be formed from these participants in each of the following cases:

 (i) *there are no restrictions?*
 (ii) *there is no unionist in the committee?*
 (iii) *the committee consists of unionists only?*
 (iv) *there is exactly one unionist in the committee?*
 (v) *there is at least one unionist in the committee?*

Solution (i) This is the number of 7-element subsets of a 19-element set. By definition, the desired number is $\binom{19}{7}$.
(ii) This is the number of ways to form a 7-member committee from the 10 non-unionists. Thus, the desired number is $\binom{10}{7}$.
(iii) Obviously, the desired number is $\binom{9}{7}$.
(iv) We first select a member from the 9 unionists and then select the remaining 6 from the 10 non-unionists. By (MP), the desired number is $\binom{9}{1}\binom{10}{6} = 9\binom{10}{6}$.

(v) There are 7 cases to consider, namely, having r unionists, where $r = 1, 2, 3, 4, 5, 6, 7$. Thus, by (AP) and (MP), the desired number is given by

$$\binom{9}{1}\binom{10}{6} + \binom{9}{2}\binom{10}{5} + \cdots + \binom{9}{6}\binom{10}{1} + \binom{9}{7}.$$

Indeed, we can have a shorter way to solve this part by using the idea of "complementation".

By (i), there are $\binom{19}{7}$ 7-member committees we can form from 19 participants. Among them, there are $\binom{10}{7}$ such committees which contain no unionist by (ii). Thus, the number of 7-member committees which contain at least one unionist should be $\binom{19}{7} - \binom{10}{7}$. (The reader may check that these two answers agree.) □

The second solution given in (v) for the above example is just an instance of applying the following principle.

Principle of Complementation (CP)

Let A be a subset of a finite set B.

Then $|B \backslash A| = |B| - |A|$, where $B \backslash A = \{x : x \in B \text{ but } x \notin A\}$.

(4.1)

If you revisit Example 2.4, you may then observe that the problem can also be solved by (CP). There are $\binom{9}{3}$ ways to form a 3-vertex subset from the given 9 vertices. Among them, the 3 on AB and any 3 on BC do not form a triangle. Thus, the number of triangles one can form is, by (CP),

$$\binom{9}{3} - \left\{\binom{3}{3} + \binom{4}{3}\right\},$$

which is 79.

We have seen from the above examples how, by applying (CP), we are able to considerably shorten the work needed to solve a counting problem. When a direct approach involves a large number of cases for which a certain condition holds, the complementary view of the *smaller* number of cases in which the condition does NOT hold allows a quicker solution to the problem. What follows then is that we count the number of ways afforded by the smaller number of "complementary" cases and finally obtain the required answer by subtracting this from the total number of ways.

Exercise

4.1 (Continuation from Example 4.1)

    (v) A and B are at the two ends;

    (vi) Z is at the centre and adjacent to A and B;

    (vii) A, B and Z form a single block (i.e. there is no other person between any two of them);

    (viii) all men form a single block;

    (ix) all men form a single block and all boys form a single block;

    (x) no two of A, B and Z are adjacent;

    (xi) all boys form a single block and Z is adjacent to A;

    (xii) Z is between A and B (need not be adjacent).

4.2 (Continuation from Example 4.2)

    (iii) the integers are odd;

    (iv) the integers are divisible by 5;

    (v) the integers are greater than 3456.

4.3 Four people can be paired off in three ways as shown below:

    (1) $\{\{A, B\}, \{C, D\}\}$,

    (2) $\{\{A, C\}, \{B, D\}\}$,

    (3) $\{\{A, D\}, \{B, C\}\}$.

    In how many ways can 10 people be paired off?

4.4 If n points on the circumference of a circle are joined by straight lines in all possible ways and no three of these lines meet at a single point inside the circle, find

    (i) the number of triangles formed with all vertices lying inside the circle;

    (ii) the number of triangles formed with exactly two vertices inside the circle;

    (iii) the number of triangles formed with exactly one vertex inside the circle;

    (iv) the total number of triangles formed.

4.5   Three girls and seven boys are to be lined up in a row. Find the number of ways this can be done if

    (i) there is no restriction;

    (ii) the girls must form a single block;

    (iii) no two girls are adjacent;

    (iv) each boy is adjacent to at most one girl.

4.6   Eight students are in a sailing club. In how many ways can they form a team consisting of 4 Laser pairs, where the order of the pairs does not matter? (Note: A Laser is a sailing boat that takes a crew of two.)

4.7   There are three boys and two girls.

    (i) Find the number of ways to arrange them in a row.

    (ii) Find the number of ways to arrange them in a row so that the two girls are next to each other.

    (iii) Find the number of ways to arrange them in a row so that there is at least one boy between the two girls.

4.8   In how many ways can a committee of 5 be formed from a group of 11 people consisting of 4 teachers and 7 students if

    (i) the committee must include exactly 2 teachers?

    (ii) the committee must include at least 3 teachers?

    (iii) a particular teacher and a particular student cannot be both in the committee?

4.9   A *palindrome* is a number that remains the same when it is read backward, for example, 2002 is a palindrome. Find the number of n-digit palindromes.

4.10   A team of 6 people is to be chosen from a list of 10 candidates. Find in how many ways this can be done

    (i) if the order of the people in the team does not matter;

    (ii) if the team consists of 6 people in a definite order;

    (iii) if the team consists of a first pair, a second pair and a third pair but order within each pair does not matter.

(C)

4.11   Find how many three figure numbers, lying between 100 and 999 inclusive, have two and only two consecutive figures identical.

(C)

4.12   Find the number of ways in which 10 persons can be divided into

    (i) two groups consisting of 7 and 3 persons;

    (ii) three groups consisting of 4, 3 and 2 persons with 1 person rejected.

                                                               (C)

4.13   (i) Find the number of integers from 100 to 500 that do not contain the digit "0".

    (ii) Find the number of integers from 100 to 500 that contain exactly one "0" as a digit.

4.14   Calculate the number of ways of selecting 2 points from 6 distinct points. Six distinct points are marked on each of two parallel lines. Calculate the number of

    (i) distinct quadrilaterals which may be formed using 4 of the 12 points as vertices;

    (ii) distinct triangles which may be formed using 3 of the 12 points as vertices.

                                                               (C)

4.15   (a) A tennis team of 4 men and 4 women is to be picked from 6 men and 7 women. Find the number of ways in which this can be done.

    (b) It was decided that 2 of the 7 women must either be selected together or not selected at all. Find how many possible teams could be selected in these circumstances. The selected team is arranged into 4 pairs, each consisting of a man and a woman. Find the number of ways in which this can be done.

                                                               (C)

Chapter 5

The Bijection Principle

We have introduced three basic principles for counting, namely, the (AP), the (MP) and the (CP). In this chapter, we shall introduce another basic principle for counting which we call the *Bijection Principle*, and discuss some of its applications.

Figure 5.1

Suppose that there are 200 parking lots in a multi-storey carpark. The carpark is full with each vehicle occupying a lot and each lot being occupied by a vehicle (see Figure 5.1). Then we know that the number of vehicles in the carpark is 200 without having to count the vehicles one by one. The number of vehicles and the number of lots are the same because there is a one to one correspondence between the set of vehicles and the set of lots in the carpark. This is a simple illustration of the Bijection Principle that we will soon state.

Let us first recall some concepts on mappings of sets. Suppose A and B are two given sets. *A mapping f from A to B*, denoted by

$$f : A \to B,$$

is a rule which assigns to each element a in A a *unique* element, denoted by $f(a)$, in B. Four examples of mappings are shown pictorially in Figure 5.2.

Certain kinds of mappings are important. Let $f : A \to B$ be a mapping. We say that f is *injective* (or *one to one*) if $f(x) \neq f(y)$ in B whenever $x \neq y$ in A. Thus, in Figure 5.2, f_2 and f_4 are injective, while

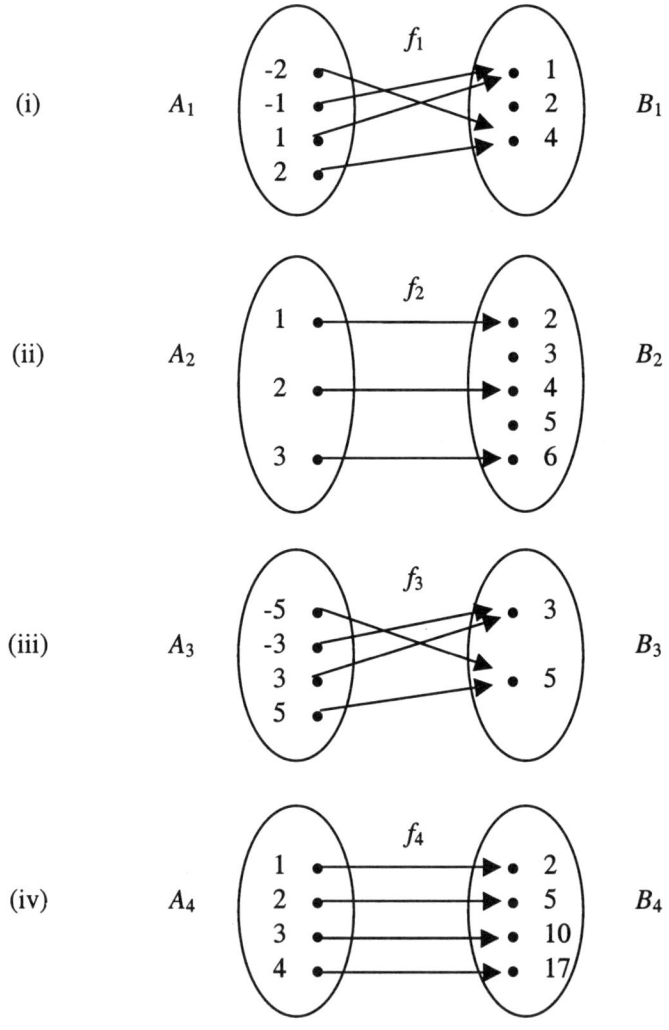

Figure 5.2

f_1 and f_3 are not (why?). We say that f is *surjective* (or *onto*) if for any b in B, there exists an a in A such that $f(a) = b$. Thus, in Figure 5.2, f_3 and f_4 are surjective whereas f_1 and f_2 are not (why?). We call f a *bijection* from A to B if f is both injective and surjective. (Sometimes, a bijection from A to B is referred to as a *one-to-one correspondence* between A and B.) Thus, in Figure 5.2, f_4 is the only bijection. These observations on the four mappings are summarized in Table 5.1.

Table 5.1

	Injection	Surjection	Bijection
f_1	x	x	x
f_2	✓	x	x
f_3	x	✓	x
f_4	✓	✓	✓

Let A and B be two finite sets. Suppose there is a mapping $f : A \to B$ that is injective. Then, by definition, each element a in A corresponds to its image $f(a)$ in B, and distinct elements in A correspond to distinct images in B. Thus, we have:

The Injection Principle (IP)
Let A and B be finite sets. If there exists a one-to-one mapping $f : A \to B$, then

$$|A| \leq |B|.$$

(5.1)

Suppose further, that the one-to-one mapping $f : A \to B$ is onto. Then each element b in B has a unique preimage a in A such that $f(a) = b$. In this case, we clearly have:

The Bijection Principle (BP)
Let A and B be finite sets. If there exists a bijection $f : A \to B$, then

$$|A| = |B|.$$

(5.2)

In this chapter, we shall focus on (BP). Through the discussions on a number of problems, we shall show you how powerful this principle is.

First of all, let us revisit a problem we studied in Chapter 4. In Example 4.5, we counted the number of chords and the number of points of intersection of the chords joining some fixed points on the circumference of a circle. Let us consider a similar problem. Figure 5.3 shows five distinct points on the circumference of a circle.

How many chords can be formed by these points?

Let A be the set of such chords, and B, the set of 2-element subsets of $\{1, 2, 3, 4, 5\}$. Given a chord α in A, define $f(\alpha) = \{p, q\}$, where p, q are the two points (on the circumference) which determine the chord α (see Figure 5.4). Then f is a mapping from A to B. Clearly, if α and β are two distinct chords in A, then $f(\alpha) \neq f(\beta)$. Thus, f is injective. On the other hand, for any 2-element subset $\{p, q\}$ in B (say, $p = 2$ and

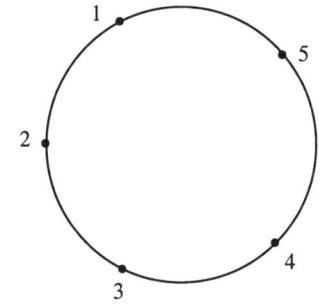

Figure 5.3

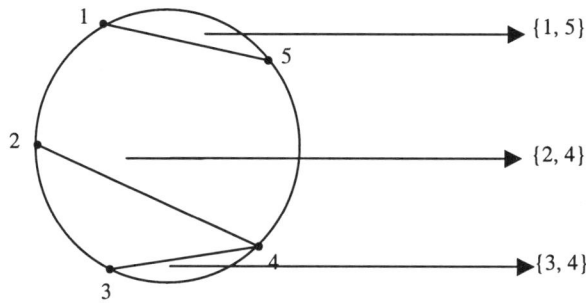

Figure 5.4

$q = 5$), there is a chord α in A (in this instance, α is the chord joining points 2 and 5) such that $f(\alpha) = \{p, q\}$. Thus, f is onto.

Hence, $f : A \to B$ is a bijection and, by (BP), we have $|A| = |B|$. As B is the set of all 2-element subsets of $\{1, 2, 3, 4, 5\}$, $|B| = \binom{5}{2}$. We thus conclude that $|A| = |B| = \binom{5}{2}$.

Next we ask: How many points of intersection (of these $\binom{5}{2}$ chords) that lie in the interior of the circle are there if no three chords are concurrent in the interior of the circle?

Let A be the set of such points of intersection and B, the set of 4-element subsets of $\{1, 2, 3, 4, 5\}$. Figure 5.5 exhibits a bijection between A and B (figure out the rule which defines the bijection!). Thus, by (BP), $|A| = |B|$. Since $|B| = \binom{5}{4}$ by definition, we have $|A| = \binom{5}{4}$.

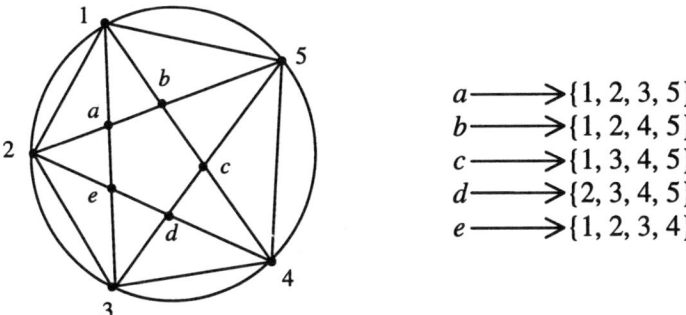

Figure 5.5

Let us proceed to show some more applications of (BP).

Example 5.1 *Figure 5.6 shows a 2×4 rectangular grid with two specified corners P and Q. There are 12 horizontal segments and 10 vertical segments in the grid. A shortest P—Q route is a continuous path from P to Q consisting of 4 horizontal segments and 2 vertical*

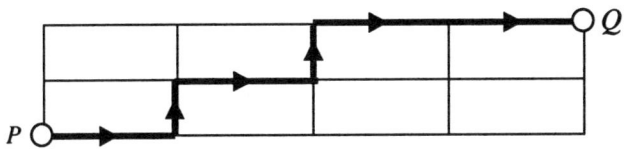

Figure 5.6

segments. An example is shown in Figure 5.6. How many shortest P—Q routes in the grid are there?

Solution   Certainly, we can solve the problem directly by listing all the possible shortest routes. This, however, would not be practical if we wish to solve the same problem in, say, a 190×100 rectangular grid. We look for a more efficient way.

There are two types of segments: horizontal and vertical. Let us use a "0" to represent a horizontal segment, and a "1" to represent a vertical segment. Thus, the shortest P—Q route shown in Figure 5.6 can accordingly be represented by the binary sequence with four "0"s and two "1"s as shown below:

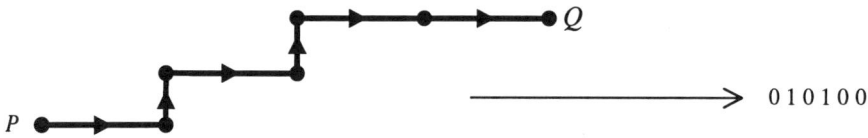

Likewise, we can have:

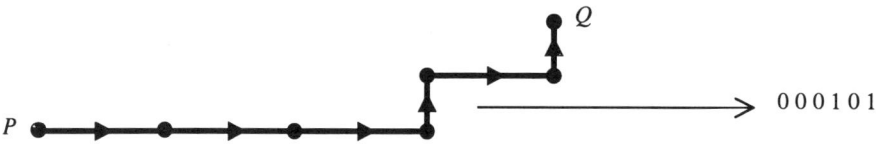

and so on.

Now, let A be the set of all shortest P—Q routes and B, the set of all 6-digit binary sequences with two 1's. Then we see that the above way of representing a shortest P—Q route in A by a binary sequence in B defines a mapping $f : A \to B$. Clearly, different shortest P—Q routes in A correspond to different sequences in B under f. Thus, f is one-to-one. Further, for any sequence b in B, say, $b = 100010$, one can find a shortest P—Q route, a in A, in this case,

so that $f(a) = b$. Thus f is onto, and so it is a bijection. Now, by (BP), we conclude that $|A| = |B|$. But how does this simplify our effort to find the number of shortest P—Q routes?

Let us explain. What is the set B? B is the set of all 6-digit binary sequences with two 1's. Can we count $|B|$? Oh, yes! We have already solved it in Example 4.4. The answer is $|B| = \binom{6}{2}$. Accordingly, we have $|A| = |B| = \binom{6}{2}$. $\qquad\qquad\qquad\qquad\qquad\square$

Example 5.2   *The power set of a set S, denoted by $\mathcal{P}(S)$, is the set of all subsets of S, inclusive of S and the empty set ϕ. Thus, for $N_n = \{1, 2, \ldots, n\}$, $1 \le n \le 3$, we have*

$$\mathcal{P}(N_1) = \{\phi, \{1\}\},$$

$$\mathcal{P}(N_2) = \{\phi, \{1\}, \{2\}, \{1, 2\}\},$$

$$\mathcal{P}(N_3) = \{\phi, \{1\}, \{2\}, \{3\}, \{1, 2\}, \{1, 3\}, \{2, 3\}, \{1, 2, 3\}\}.$$

Note that $|\mathcal{P}(N_1)| = 2, |\mathcal{P}(N_2)| = 4, |\mathcal{P}(N_3)| = 8$. Table 3.1 shows that $|\mathcal{P}(N_4)| = 16$. What is the value of $|\mathcal{P}(N_5)|$?

Solution   For convenience, let $A = \mathcal{P}(N_5)$; that is, A is the power set of $\{1, 2, 3, 4, 5\}$. Represent these subsets by 5-digit binary sequences as follows:

$$\phi \qquad \longrightarrow 00000$$
$$\{1\} \qquad \longrightarrow 10000$$
$$\{2\} \qquad \longrightarrow 01000$$
$$\vdots$$
$$\{5\} \qquad \longrightarrow 00001$$
$$\{1, 2\} \qquad \longrightarrow 11000$$
$$\vdots$$
$$\{4, 5\} \qquad \longrightarrow 00011$$
$$\vdots$$
$$\{1, 3, 5\} \qquad \longrightarrow 10101$$
$$\vdots$$
$$\{1, 2, 3, 4, 5\} \longrightarrow 11111$$

The rule is that the ith digit of the corresponding binary sequence is "1" if "i" is in the subset; and "0" otherwise. Let B be the set of all 5-digit binary sequences. Clearly, the above rule establishes a bijection between A and B. Thus, by (BP), $|A| = |B|$. Since $|B| = 2^5$ (see Example 2.2), $|A| = 2^5$.   □

Note that $|\mathcal{P}(\text{N}_1)| = 2 = 2^1$, $|\mathcal{P}(\text{N}_2)| = 4 = 2^2$, $|\mathcal{P}(\text{N}_3)| = 8 = 2^3$, $|\mathcal{P}(\text{N}_4)| = 16 = 2^4$, and now $|\mathcal{P}(\text{N}_5)| = 2^5$. What is $|\mathcal{P}(\text{N}_n)|$ for $n \geq 1$? (See Exercise 5.3.)

Finally, let us introduce a counting problem related to the notion of divisors of natural numbers. We shall denote by N, the set of natural numbers; i.e.

$$\text{N} = \{1, 2, 3, \ldots\}.$$

Assume that $d, n \in \text{N}$. We say that d is a *divisor* of n if when n is divided by d, the remainder is zero. Thus, 3 is a divisor of 12, 5 is a divisor of 100, but 2 is not a divisor of 9.

Let $n \in \text{N}$, $n \geq 2$. Clearly, n has at least two divisors, namely 1 and n. How many divisors (inclusive of 1 and n) does n have? This is a type of problem that can often be found in mathematical competitions. We shall tackle this problem and see how (MP) and (BP) are used in solving the problem.

To understand the solution, we first recall a special type of numbers called prime numbers and state an important result relating natural numbers and prime numbers.

A natural number p is said to be *prime* (or called a *prime*) if $p \geq 2$ and the only divisors of p are 1 and p. All prime numbers less than 100 are shown below:

$$2, 3, 5, 7, 11, 13, 17, 19, 23, 29, 31, 37, 41,$$
$$43, 47, 53, 59, 61, 67, 71, 73, 79, 83, 89, 97.$$

The primes are often referred to as building blocks of numbers because every natural number can always be expressed uniquely as a product of some primes. For example,

$$108 = 2^2 \times 3^3, \qquad\qquad 1620 = 2^2 \times 3^4 \times 5,$$
$$1815 = 3 \times 5 \times 11^2, \qquad 215306 = 2 \times 7^2 \times 13^3.$$

This fact is so basic and important to the study of numbers that it is called the *Fundamental Theorem of Arithmetic* (FTA).

(FTA)   Every natural number $n \geq 2$ can be factorized as

$$n = p_1^{m_1} p_2^{m_2} \cdots p_k^{m_k}$$

for some distinct primes $p_1, p_2, \ldots, p_k$ and for some natural numbers $m_1, m_2, \ldots, m_k$. Such a factorization is unique if the order of primes is disregarded.

FTA was first studied by the Greek mathematician, Euclid (c. 450–380 BC) in the case where the number of primes is at most 4. It was the German mathematician, Carl Friedrich Gauss (1777–1855), known as the Prince of Mathematicians, who stated and proved the full result in 1801.

Let us now return to the problem of counting the number of divisors of n. How many divisors does the number 72 have? Since 72 is not a big number, we can get the answer simply by listing all the divisors of 72:

$$1, 2, 3, 4, 6, 8, 9, 12, 18, 24, 36, 72 \,.$$

The way of counting the divisors of n by listing as shown above is certainly impractical when n gets larger. We look for a more efficient way.

Let us look at the example when $n = 72$ again and try to get some information about 72 and its divisors by FTA.

The images above are those of Euclid on a stamp of the Maldives and Gauss on a German banknote.

Observe that $72 = 2^3 \times 3^2$. Suppose x is a divisor of 72. Clearly, x does not contain prime factors other than 2 and 3. That is, x must be of the form

$$x = 2^p \times 3^q$$

where, clearly, $p \in \{0, 1, 2, 3\}$ and $q \in \{0, 1, 2\}$. On the other hand, any such number $2^p \times 3^q$ is a divisor of 72. Indeed,

$$1 = 2^0 \times 3^0, \qquad 9 = 2^0 \times 3^2,$$
$$2 = 2^1 \times 3^0, \qquad 12 = 2^2 \times 3^1,$$
$$3 = 2^0 \times 3^1, \qquad 18 = 2^1 \times 3^2,$$
$$4 = 2^2 \times 3^0, \qquad 24 = 2^3 \times 3^1,$$
$$6 = 2^1 \times 3^1, \qquad 36 = 2^2 \times 3^2,$$
$$8 = 2^3 \times 3^0, \qquad 72 = 2^3 \times 3^2.$$

Let A be the set of divisors of 72 and $B = \{(p, q) : 0 \le p \le 3, 0 \le q \le 2\} = \{0, 1, 2, 3\} \times \{0, 1, 2\}$. Then the above list implies that the mapping f defined by

$$f(1) = (0, 0), \qquad f(9) = (0, 2),$$
$$f(2) = (1, 0), \qquad f(12) = (2, 1),$$
$$f(3) = (0, 1), \qquad f(18) = (1, 2),$$
$$f(4) = (2, 0), \qquad f(24) = (3, 1),$$
$$f(6) = (1, 1), \qquad f(36) = (2, 2),$$
$$f(8) = (3, 0), \qquad f(72) = (3, 2),$$

is a bijection from A to B. Thus, by (BP) and (MP), $|A| = |B| = |\{0, 1, 2, 3\} \times \{0, 1, 2\}| = |\{0, 1, 2, 3\}| \times |\{0, 1, 2\}| = 4 \times 3 = 12$, which agrees with the above listing.

The following example extends what we discussed above.

Example 5.3 *Find the number of divisors of* 12600.

Solution Observe that $12600 = 2^3 \times 3^2 \times 5^2 \times 7^1$.

Thus a number z is a divisor of 12600 if and only if it is of the form

$$z = 2^a \times 3^b \times 5^c \times 7^d$$

where a, b, c, d are integers such that $0 \leq a \leq 3, 0 \leq b \leq 2, 0 \leq c \leq 2$ and $0 \leq d \leq 1$.

Let A be the set of divisors z of 12600 and $B = \{(a, b, c, d) : a = 0, 1, 2, 3; b = 0, 1, 2; c = 0, 1, 2; d = 0, 1\}$. Clearly, the mapping f defined by

$$f(z) = (a, b, c, d),$$

is a bijection from A to B. Then, by (BP) and (MP), $|A| = |B| = 4 \cdot 3 \cdot 3 \cdot 2 = 72.$ $\square$

We have seen from the above examples how crucial applying (BP) is as a step towards solving a counting problem. Given a finite set A, the objective is to enumerate $|A|$, but of course, this is not easy. In the course of applying (BP), we look for a more familiar finite set B and try to establish a bijection between these two sets. Once this is done, the harder problem of counting $|A|$ is transformed to an easier problem (hopefully) of counting $|B|$. It does not matter how different the members in A and those in B are in nature. As long as there exists a bijection between them, we get $|A| = |B|$.

Exercise

5.1 (a) Find the number of positive divisors of n if

  (i) $n = 31752$;
  (ii) $n = 55125$.

  (b) In general, given an integer $n \geq 2$, how do you find the number of positive divisors of n?

5.2 In each of the following cases, find the number of shortest P—Q routes in the grid below:

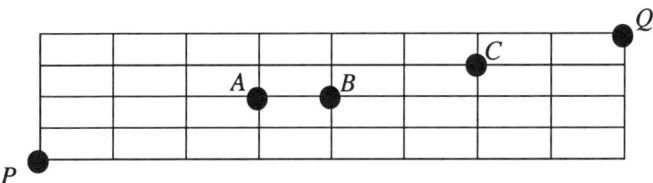

(i) the routes must pass through A;

(ii) the routes must pass through AB;

(iii) the routes must pass through A and C;

(iv) the segment AB is deleted.

5.3 For each positive integer n, show that $|\mathcal{P}(N_n)| = 2^n$ by establishing a bijection between $\mathcal{P}(N_n)$ and the set of n-digit binary sequences.

5.4 Let n and r be integers with $1 \leq r \leq n$. Prove that $\binom{n}{r} = \binom{n}{n-r}$ by establishing a bijection between the set of r-element subsets of N_n and the set of $(n-r)$-element subsets of N_n.

5.5 The number 4 can be expressed as a sum of one or more positive integers, taking order into account, in the following 8 ways:

$$
\begin{aligned}
4 = 4 &= 1 + 3 \\
&= 3 + 1 = 2 + 2 \\
&= 1 + 1 + 2 = 1 + 2 + 1 \\
&= 2 + 1 + 1 = 1 + 1 + 1 + 1 .
\end{aligned}
$$

Show that every natural number n can be so expressed in 2^{n-1} ways.

5.6 How many rectangles are there in the following 6×7 grid?

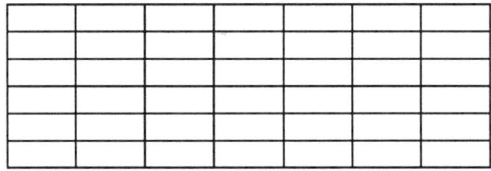

5.7 Find the number of parallelograms which are contained in the configuration below and which have no sides parallel to BC. (Hint: Adjoin a new row at the base of the triangle.)

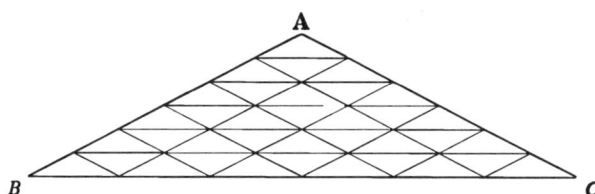

Chapter 6

Distribution of Balls into Boxes

Figure 6.1 shows three distinct boxes into which seven identical (indistinguishable) balls are to be distributed. Three different ways of distribution are shown in Figure 6.2. (Note that the two vertical bars at the two ends are removed.)

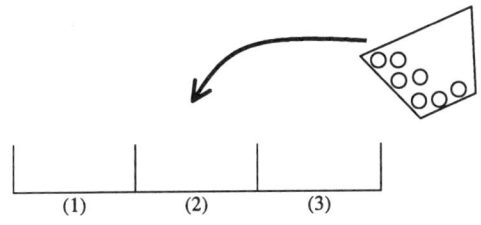

Figure 6.1

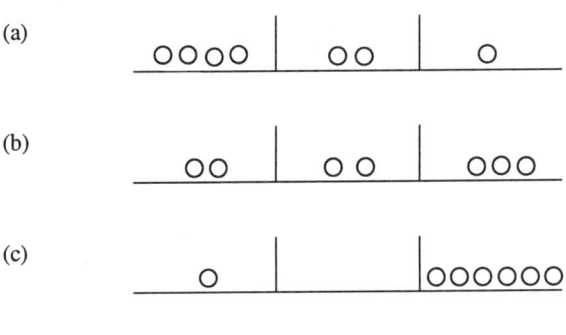

Figure 6.2

In how many different ways can this be done? This is an example of the type of problem we shall discuss in this chapter. We shall see how problems of this type can be solved by applying (BP).

In Figure 6.2, by treating each vertical bar as a "1" and each ball as a "0", each way of distribution becomes a 9-digit binary sequence with two 1's. For instance,

$$\text{(a)} \longrightarrow 0\,0\,0\,0\,1\,0\,0\,1\,0,$$
$$\text{(b)} \longrightarrow 0\,0\,1\,0\,0\,1\,0\,0\,0,$$
$$\text{(c)} \longrightarrow 0\,1\,1\,0\,0\,0\,0\,0\,0.$$

Obviously, this correspondence establishes a bijection between the set of ways of distributing the balls and the set of 9-digit binary sequences with two 1's. Thus, by (BP), the number of ways of distributing the seven identical balls into three distinct boxes is $\binom{9}{2}$.

In general, we have:

The number of ways of distributing r *identical* balls into n *distinct* boxes is given by $\binom{r+n-1}{n-1}$, which is equal to $\binom{r+n-1}{r}$, by (3.6).   (6.1)

In the distribution problem discussed above, some boxes may be vacant at the end. Supposing no box is allowed to be vacant, how many ways are there to distribute the seven identical balls into three distinct boxes?

To meet the requirement that no box is vacant, we first put a ball in each box and this is counted as one way because the balls are identical. We are then left with $4\,(=7-3)$ balls, but we are now free to distribute these 4 balls into any box. By the result (6.1), the number of ways this can be done is $\binom{4+3-1}{3-1} = \binom{6}{2}$. Thus, the number of ways to distribute 7 identical balls into 3 distinct boxes so that no box is empty is $\binom{6}{2}$.

In general, suppose we wish to distribute r *identical* balls into n *distinct* boxes, where $r \geq n$, in such a way that no box is vacant. This can be done in the following steps: First, we put one ball in each box; and then distribute the remaining $r - n$ balls to the n boxes in any arbitrary way. As the balls are identical, the number of ways for the first step to be done is 1. On the other hand, by the result (6.1), the number of ways to do the second step is

$$\binom{(r-n)+n-1}{n-1}.$$

Thus, by (MP) and upon simplification, we arrive at the following result.

The number of ways to distribute r *identical* balls into n *distinct* boxes, where $r \geq n$, so that no box is empty is given by $\binom{r-1}{n-1}$, which is equal to $\binom{r-1}{r-n}$.   (6.2)

Example 6.1 *There are 11 men waiting for their turn in a barber shop. Three particular men are A, B and C. There is a row of 11 seats for the customers. Find the number of ways of arranging them so that no two of A, B and C are adjacent.*

Solution There are different ways to solve this problem. We shall see in what follows that it can be treated as a distribution problem.

First of all, there are 3! ways to arrange A, B and C. Fix one of the ways, say A—B—C. We then consider the remaining 8 persons. Let us imagine tentatively that these 8 persons are identical, and they are to be placed in 4 distinct boxes as shown in Figure 6.3 so that boxes (2) and (3) are not vacant (since no two of A, B and C are adjacent). To meet this requirement, we place one in box (2) and one in box (3). Then the remaining six can be placed freely in the boxes in $\binom{6+4-1}{4-1} = \binom{9}{3}$ ways by (6.1). (Figure 6.4 shows a way of distribution.)

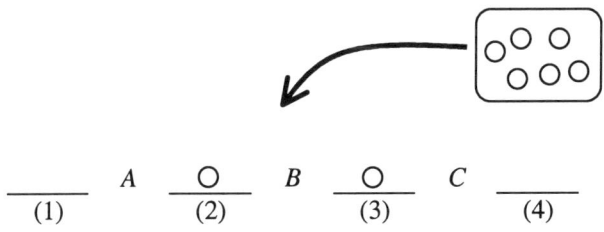

$$\underline{\hspace{2cm}} \quad A \quad \underset{(2)}{\bigcirc} \quad B \quad \underset{(3)}{\bigcirc} \quad C \quad \underline{\hspace{2cm}}$$
$$\quad (1) \qquad\qquad\qquad\qquad\qquad\qquad\qquad (4)$$

Figure 6.3

$$\underline{\underset{(1)}{\bigcirc\bigcirc}} \quad A \quad \underline{\underset{(2)}{\bigcirc\bigcirc}} \quad B \quad \underline{\underset{(3)}{\bigcirc}} \quad C \quad \underline{\underset{(4)}{\bigcirc\bigcirc\bigcirc}}$$

Figure 6.4

But the eight persons are actually distinct. Thus, to each of these $\binom{9}{3}$ ways, there are 8! ways to arrange them.

Hence by (MP), the required number of ways is $3!\binom{9}{3}8!$, which is $8!\,9 \cdot 8 \cdot 7$. $\qquad\qquad\qquad\square$

Remark   The answer, $8!\,9 \cdot 8 \cdot 7$, suggests that the problem can be solved in the following way. We first arrange the 8 persons (excluding A, B and C) in a row in 8! ways. Fix one of these ways, say

$$\underline{\quad}\underset{(1)}{}\ X_1\ \underline{\quad}\underset{(2)}{}\ X_2\ \underline{\quad}\underset{(3)}{}\ X_3\ \underline{\quad}\underset{(4)}{}\ X_4\ \underline{\quad}\underset{(5)}{}\ X_5\ \underline{\quad}\underset{(6)}{}\ X_6\ \underline{\quad}\underset{(7)}{}\ X_7\ \underline{\quad}\underset{(8)}{}\ X_8\ \underline{\quad}\underset{(9)}{}$$

We now consider A. There are 9 ways to place A in one of the 9 boxes, say box (4):

$$\underline{\quad}\underset{(1)}{}\ X_1\ \underline{\quad}\underset{(2)}{}\ X_2\ \underline{\quad}\underset{(3)}{}\ X_3\ \underline{\quad}\underset{(4)}{}\ \overset{A}{}\ X_4\ \underline{\quad}\underset{(5)}{}\ X_5\ \underline{\quad}\underset{(6)}{}\ X_6\ \underline{\quad}\underset{(7)}{}\ X_7\ \underline{\quad}\underset{(8)}{}\ X_8\ \underline{\quad}\underset{(9)}{}$$

Next, consider B. Since A and B cannot be adjacent, B can be placed only in one of the remaining 8 boxes. Likewise, C can be placed only in one of the remaining 7 boxes. The answer is thus $8!\,9 \cdot 8 \cdot 7$.

Exercise

6.1   There are four types of sandwiches. A boy wishes to place an order of 3 sandwiches. How many such orders can he place?

6.2   Calculate the number of distinct 9-letter arrangements which can be made with letters of the word SINGAPORE such that no two vowels are adjacent.

6.3   There is a group of 10 students which includes three particular students A, B and C. Find the number of ways of arranging the 10 students in a row so that B is always between A and C. (A and B, or B and C need not be adjacent.)

6.4 Six distinct symbols are transmitted through a communication channel. A total of 18 blanks are to be inserted between the symbols with at least 2 blanks between every pair of symbols. In how many ways can the symbols and blanks be arranged?

Chapter 7

More Applications of (BP)

We shall give additional examples in this chapter to show more applications of (BP).

Consider the following linear equation:

$$x_1 + x_2 + x_3 = 7. \qquad (1)$$

If we put $x_1 = 4, x_2 = 1$ and $x_3 = 2$, we see that (1) holds. Since 4, 1, 2 are nonnegative integers, we say that $(x_1, x_2, x_3) = (4, 1, 2)$ is a *nonnegative integer solution* to the linear equation (1). Note that $(x_1, x_2, x_3) = (1, 2, 4)$ is also a nonnegative integer solution to (1), and so are (4, 2, 1) and (1, 4, 2). Other nonnegative integer solutions to (1) include

$$(0, 0, 7), (0, 7, 0), (1, 6, 0), (5, 1, 1), \dots .$$

Example 7.1 *Find the number of nonnegative integer solutions to* (1).

Solution Let us create 3 distinct "boxes" to represent x_1, x_2 and x_3, respectively. Then each nonnegative integer solution $(x_1, x_2, x_3) = (a, b, c)$ to (1) corresponds, in a natural way, to a way of distributing 7 identical balls into boxes so that there are a, b and c balls in boxes (1), (2) and (3) respectively (see Figure 7.1).

This correspondence clearly establishes a bijection between the set of nonnegative integer solutions to (1) and the set of ways of distributing 7 identical balls in 3 distinct boxes. Thus, by (BP) and the result of (6.1), the number of nonnegative integer solutions to (1) is $\binom{7+3-1}{3-1} = \binom{9}{2}$. $\square$

$$(4, 1, 2) \quad \longrightarrow \quad \underline{0000 \mid 0 \mid 00}$$
$$\qquad\qquad\qquad\qquad (1) \quad (2) \quad (3)$$

$$(2, 5, 0) \quad \longrightarrow \quad \underline{00 \mid 00000 \mid}$$
$$\qquad\qquad\qquad\qquad (1) \quad (2) \quad\;\; (3)$$

Figure 7.1

By generalizing the above argument and applying the results (6.1) and (6.2), we can actually establish the following general results.

Consider the linear equation

$$x_1 + x_2 + \cdots + x_n = r \qquad\qquad (2)$$

where r is a nonnegative integer.

(i) The number of nonnegative integer solutions to (2) is given by $\binom{r+n-1}{r}$.

(ii) The number of positive integer solutions $(x_1, x_2, \ldots, x_n)$ to (2), with each $x_i \geq 1$, is given by $\binom{r-1}{r-n}$, where $r \geq n$ and $i = 1, 2, \ldots, n$.

(7.1)

Example 7.2 *Recall that the number of 3-element subsets $\{a, b, c\}$ of the set $N_{10} = \{1, 2, 3, \ldots, 10\}$ is $\binom{10}{3}$. Assume that $a < b < c$ and suppose further that*

$$b - a \geq 2 \quad \text{and} \quad c - b \geq 2 \qquad\qquad (3)$$

(i.e. no two numbers in $\{a, b, c\}$ are consecutive). For instance, $\{1, 3, 8\}$ and $\{3, 6, 10\}$ satisfy (3) but not $\{4, 6, 7\}$ and $\{1, 2, 9\}$. How many such 3-element subsets of N_{10} are there?

Solution Let us represent a 3-element subset $\{a, b, c\}$ of N_{10} satisfying (3) by a 10-digit binary sequence as follows:

	(1)	(2)	(3)	(4)	(5)	(6)	(7)	(8)	(9)	(10)
$\{1, 3, 8\} \longrightarrow$	1	0	1	0	0	0	0	1	0	0
$\{3, 6, 10\} \longrightarrow$	0	0	1	0	0	1	0	0	0	1

Note that the rule is similar to the one introduced in Example 5.2. Clearly, this correspondence is a bijection between the set A of 3-element subsets of N_{10} satisfying (3) and the set B of 10-digit binary sequences with three 1's in which no two 1's are adjacent. Thus $|A| = |B|$. But how do we count $|B|$? Using the method discussed in Example 6.1, we obtain

$$|B| = \binom{(7-2)+4-1}{4-1} = \binom{8}{3}.$$

Thus $|A| = \binom{8}{3}$. □

Example 7.3 *Two tennis teams A and B, consisting of 5 players each, will have a friendly match playing only singles tennis with no ties allowed. The players in each team are arranged in order:*

$$A : a_1, a_2, a_3, a_4, a_5,$$
$$B : b_1, b_2, b_3, b_4, b_5.$$

The match is run in the following way. First, a_1 plays against b_1. Suppose a_1 wins (i.e. b_1 is eliminated). Then a_1 continues to play against b_2; if a_1 is beaten by b_2 (i.e. a_1 is eliminated), then b_2 continues to play against a_2, and so on. What is the number of possible ways in which all the 5 players in team B are eliminated? (Two such ways are shown in Figure 7.2.)

Solution Let x_i be the number of games won by player a_i, $i = 1, 2, 3, 4, 5$. Thus, in Figure 7.2(i),

$$x_1 = 2, \quad x_2 = 0, \quad x_3 = 3, \quad x_4 = x_5 = 0$$

and in Figure 7.2(ii),

$$x_1 = x_2 = 0, \quad x_3 = 2, \quad x_4 = 1, \quad x_5 = 2.$$

In order for the 5 players in team B to be eliminated, we must have

$$x_1 + x_2 + x_3 + x_4 + x_5 = 5 \tag{4}$$

and the number of ways this can happen is, by (BP), the number of nonnegative integer solutions to (4). Thus, the desired answer is $\binom{5+5-1}{4} = \binom{9}{4}$, by the result 7.1(i). □

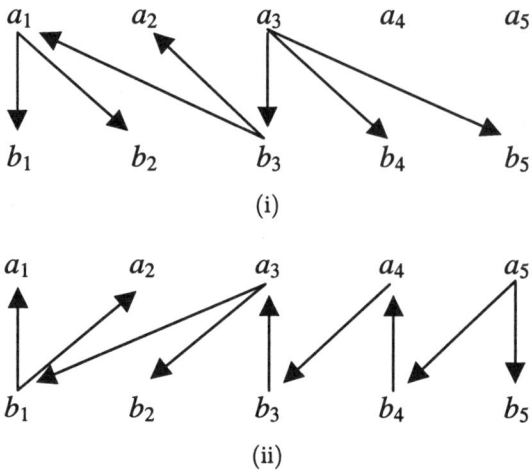

(i)

(ii)

"$a \to b$" means "a beats b"

Figure 7.2

Example 7.4 *Eight letters are to be selected from the five vowels a, e, i, o, u with repetition allowed. In how many ways can this be done if*

(i) *there are no other restrictions?*
(ii) *each vowel must be selected at least once?*

Solution (i) Some examples of ways of the selection are given below:

(1) a, a, u, u, u, u, u, u;
(2) a, e, i, i, i, o, o, u;
(3) e, e, i, i, o, o, u, u.

As shown in Figure 7.3, these selections can be treated as ways of distributing 8 identical objects into 5 distinct boxes.

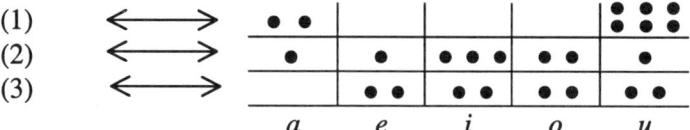

Figure 7.3

Thus, by (BP) and the result (6.1), the number of ways of selection is given by $\binom{8+5-1}{8}$, i.e. $\binom{12}{4}$.

(ii) As shown in the second row of Figure 7.3, a way of selection which includes each vowel can be treated as a way of distribution such that no box is empty. Thus, by (BP) and the result (6.2), the number of ways of selection is given by $\binom{8-1}{8-5}$, i.e. $\binom{7}{3}$. $\qquad\square$

Example 7.5 *Consider the following two 13-digit binary sequences:*

$$1\,1\,1\,0\,1\,0\,1\,1\,1\,0\,0\,0\,0,$$
$$1\,0\,0\,0\,1\,1\,0\,0\,1\,1\,1\,1\,0.$$

For binary sequences, any block of two adjacent digits is of the form $00, 01, 10$ *or* 11. *In each of the above sequences, there are three* 00, *two* 01, *three* 10 *and four* 11. *Find the number of 13-digit binary sequences which have exactly three* 00, *two* 01, *three* 10 *and four* 11.

Solution To have exactly three 10 and two 01 in a sequence, such a sequence must begin with 1, end with 0, and have the changeovers of 1 and 0 as shown below, where each of the boxes (1), (3) and (5) (respectively (2), (4) and (6)) contains only 1's (respectively 0's) and at least one 1 (respectively 0).

	[10]		[01]		[10]		[01]		[10]	
1		0		1		0		1		0
(1)		(2)		(3)		(4)		(5)		(6)

For instance, the two sequences given in the problem are of the form:

111	0	1	0	111	0000
(1)	(2)	(3)	(4)	(5)	(6)

1	000	11	00	1111	0
(1)	(2)	(3)	(4)	(5)	(6)

To have three 00 and four 11 in such a sequence, we must

(i) put in three more 0's in boxes (2), (4) or (6) (but in an arbitrary way), and

(ii) put four more 1's in boxes (1), (3) or (5) (also in an arbitrary way).

(Check that there are 13 digits altogether.) The number of ways to do (i) is $\binom{3+3-1}{3}$, i.e. $\binom{5}{2}$; while that of (ii) is $\binom{4+3-1}{4}$, i.e. $\binom{6}{2}$. Thus, by (MP), the number of such sequences is $\binom{5}{2}\binom{6}{2}$, i.e. 150. □

Example 7.6 *Consider the following three arrangements of 5 persons A, B, C, D, E in a circle:*

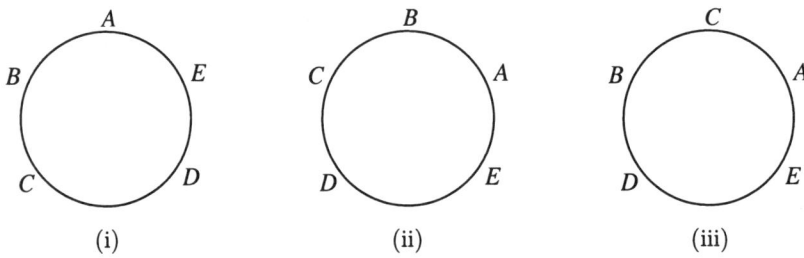

Figure 7.4

Two arrangements of n objects in a circle are considered different if and only if there is at least one object whose neighbour on the right is different in the two arrangements. Thus arrangements (i) and (ii) above are considered identical, while arrangement (iii) is considered different from (i) and (ii). (Note that the right neighbour of A in arrangement (iii) is C while that in both (i) and (ii) is B.) Find the number of arrangements of the 5 persons in a circle.

Solution For each arrangement of the 5 persons in a circle, let us line the 5 in a row as follows: We always start with A at the left end. Then we place the right neighbour of A (in the circle) to the right of A in the row. We continue, in turn, to place the right neighbour (in the circle) of the last placed person to his right in the row until every person is arranged in the row. (We can also visualize this as cutting the circle at A and then unraveling it to form a line.) Then each circular arrangement of the 5 persons corresponds to an arrangement of 5 persons in a row with A at the left end. Now, since A is always fixed at the left end, he can be neglected and the arrangement of 5 persons in a row can be seen to correspond to an arrangement of only 4 persons (B, C, D, E) in a row (see Figure 7.5).

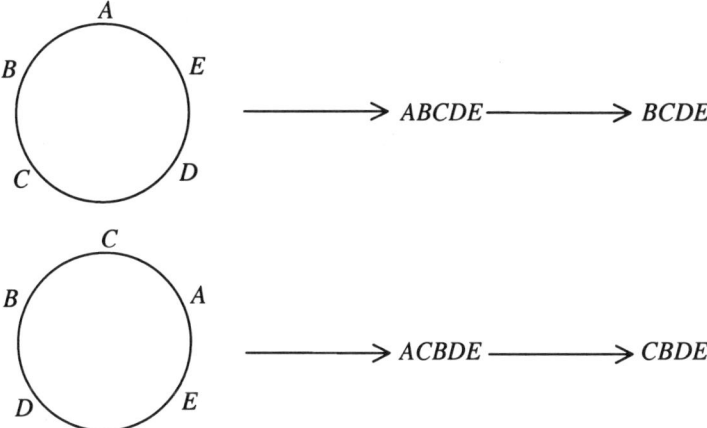

Figure 7.5

This correspondence clearly establishes a bijection between the set of arrangements of 5 persons in a circle and the set of arrangements of 4 persons in a row. Thus, by (BP) and the result of (3.1), the number of arrangements of 5 persons in a circle is 4!. □

By generalizing the above argument, we can establish the following result:

The number of ways of arranging n distinct objects in a circle is given by $(n-1)!$.

$$(7.2)$$

Exercise

7.1 Find the number of integer solutions to the equation:

$$x_1 + x_2 + x_3 + x_4 + x_5 = 51$$

in each of the following cases:

(i) $x_i \geq 0$ for each $i = 1, 2, \ldots, 5$;

(ii) $x_1 \geq 3, x_2 \geq 5$ and $x_i \geq 0$ for each $i = 3, 4, 5$;

(iii) $0 \leq x_1 \leq 8$ and $x_i \geq 0$ for each $i = 2, 3, 4, 5$;

(iv) $x_1 + x_2 = 10$ and $x_i \geq 0$ for each $i = 1, 2, \ldots, 5$;

(v) x_i is positive and odd (respectively, even) for each $i = 1, 2, \ldots, 5$.

7.2 An illegal gambling den has 8 rooms, each named after a different animal. The gambling lord needs to distribute 16 tables into the rooms. Find the number of ways of distributing the tables into the rooms in each of the following cases:

(i) Horse Room holds at most 3 tables.

(ii) Each of Monkey Room and Tiger Room holds at least 2 tables.

7.3 The number 6 can be expressed as a product of three factors in 9 ways as follows:

$$1 \cdot 1 \cdot 6, \quad 1 \cdot 6 \cdot 1, \quad 6 \cdot 1 \cdot 1, \quad 1 \cdot 2 \cdot 3, \quad 1 \cdot 3 \cdot 2, \quad 2 \cdot 1 \cdot 3, \quad 2 \cdot 3 \cdot 1, \quad 3 \cdot 1 \cdot 2, \quad 3 \cdot 2 \cdot 1.$$

In how many ways can each of the following numbers be so expressed?

(i) 2592

(ii) 27000

7.4 Find the number of integer solutions to the equation:

$$x_1 + x_2 + x_3 + x_4 = 30$$

in each of the following cases:

(i) $x_i \geq 0$ for each $i = 1, 2, 3, 4$;

(ii) $2 \leq x_1 \leq 7$ and $x_i \geq 0$ for each $i = 2, 3, 4$;

(iii) $x_1 \geq -5, x_2 \geq -1, x_3 \geq 1$ and $x_4 \geq 2$.

7.5 Find the number of quadruples (w, x, y, z) of nonnegative integers which satisfy the inequality

$$w + x + y + z \leq 2002.$$

7.6 Find the number of nonnegative integer solutions to the equation:

$$5x_1 + x_2 + x_3 + x_4 = 14.$$

7.7 There are five ways to express 4 as a sum of two nonnegative integers in which the order matters:

$$4 = 4 + 0 = 3 + 1 = 2 + 2 = 1 + 3 = 0 + 4.$$

Given $r, n \in \mathbb{N}$, what is the number of ways to express r as a sum of n nonnegative integers in which the order matters?

7.8 There are six ways to express 5 as a sum of three positive integers in which the order matters:

$$5 = 3+1+1 = 2+2+1 = 2+1+2 = 1+3+1 = 1+2+2 = 1+1+3.$$

Given $r, n \in \mathbb{N}$ with $r \geq n$, what is the number of ways to express r as a sum of n positive integers in which the order matters?

7.9 Find the number of 4-element subsets $\{a, b, c, d\}$ of the set $\mathbb{N}_{20} = \{1, 2, \ldots, 20\}$ satisfying the following condition

$$b - a \geq 2, \quad c - b \geq 3 \quad \text{and} \quad d - c \geq 4.$$

7.10 In a sequence of coin tosses, one can keep a record of the number of instances when a tail is immediately followed by a head, a head is immediately followed by a head, etc. We denote these by TH, HH, etc. For example, in the sequence $HHTTHHHHTHHTTTT$ of 15 coin tosses, we observe that there are five HH, three HT, two TH and four TT subsequences. How many different sequences of 15 coin tosses will contain exactly two HH, three HT, four TH and five TT subsequences?

(AIME)

7.11 Show that the number of ways of distributing r identical objects into n distinct boxes such that Box 1 can hold at most one object is given by

$$\binom{r+n-3}{r-1} + \binom{r+n-2}{r}.$$

7.12 In a new dictatorship, it is decided to reorder the days of the week using the same names of the days. All the possible ways of doing so are to be presented to the dictator for her to decide on one. How many ways are there in which Sunday is immediately after Friday and immediately before Thursday?

7.13 Five couples occupy a round table at a wedding dinner. Find the number of ways for them to be seated if:

(i) every man is seated between two women;

    (ii) every man is seated between two women, one of whom is his wife;

    (iii) every man is seated with his wife;

    (iv) the women are seated on consecutive seats.

7.14   The seats at a round table are numbered from 1 to 7. Find the number of ways in which a committee consisting of four men and three women can be seated at the table

    (i) if there are no restrictions;

    (ii) if all the men sit together.

                                                                          (C)

7.15   Four men, two women and a child sit at a round table. Find the number of ways of arranging the seven people if the child is seated

    (i) between the two women;

    (ii) between two men.

                                                                          (C)

Chapter 8

Distribution of Distinct Objects into Distinct Boxes

We have seen from the various examples given in Chapters 6 and 7 that the *distribution problem*, which deals with the counting of ways of distributing objects into boxes, is a basic model for many counting problems. In distribution problems, objects can be identical or distinct, and boxes too can be identical or distinct. Thus, there are, in general, four cases to be considered, namely

Table 8.1

	Objects	Boxes
(1)	identical	distinct
(2)	distinct	distinct
(3)	distinct	identical
(4)	identical	identical

We have considered Case (1) in Chapters 6 and 7. Cases (3) and (4) will not be touched upon in this booklet. In this chapter, we shall consider Case (2).

Suppose that 5 distinct balls are to be put into 7 distinct boxes.

Example 8.1 In how many ways can this be done if each box can hold at most one ball?

Example 8.2 In how many ways can this be done if each box can hold any number of balls?

Solution Before we proceed, we would like to point out that the *ordering* of the distinct objects in each box is *not* taken into consideration in the discussion in this chapter.

We first consider Example 8.1. As shown in Figure 8.1, let a, b, c, d and e denote the 5 distinct balls. First, we put a (say) into one of the boxes. There are 7 choices.

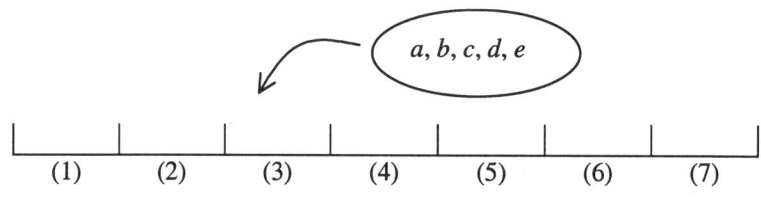

Figure 8.1

Next, we consider b (say). As each box can hold at most one ball, and one of the boxes is occupied by a, there are now 6 choices for b. Likewise, there are, respectively, 5, 4 and 3 choices for c, d and e. Thus, by (MP), the number of ways of distribution is given by $7 \cdot 6 \cdot 5 \cdot 4 \cdot 3$.

Note that the above answer can be expressed as P_5^7 which, as defined in Chapter 3, is the number of ways of arranging any 5 objects from 7 distinct objects. The fact that the above answer is P_5^7 does not surprise us as there is a 1–1 correspondence between the distributions of 5 distinct balls into 7 distinct boxes and the arrangements of 5 distinct objects from 7 distinct objects as shown in Figure 8.2. (Find out the rule of the correspondence.)

$\{a, b, c, d, e\}$ $\{1, 2, 3, 4, 5, 6, 7\}$

b	c		a	e		d	$\longleftrightarrow$	41275
e	d	c	b			a	$\longleftrightarrow$	74321
(1)	(2)	(3)	(4)	(5)	(6)	(7)		

Figure 8.2

In general, we have:

> The number of ways of distributing r distinct objects into n distinct boxes such that each box can hold at most one object (and thus $r \le n$) is given by P_r^n, which is equal to $n!/(n-r)!$.

(8.1)

We now consider Example 8.2. There are 7 ways of putting a in the boxes. As each box can hold any number of balls, there are also 7 choices for each of the remaining balls b, c, d and e. Thus, by (MP), the answer is 7^5.

In general, we have:

> The number of ways of distributing r distinct objects into n distinct boxes such that each box can hold any number of objects is given by n^r.

(8.2)

Exercise

8.1 Find the number of ways for a teacher to distribute 6 different books to 9 students if

    (i) there is no restriction;

    (ii) no student gets more than one book.

8.2 Let A be the set of ways of distributing 5 distinct objects into 7 distinct boxes with no restriction, and let B be the set of 5-digit numbers using 1, 2, 3, 4, 5, 6, 7 as digits with repetition allowed. Establish a bijection between A and B.

8.3 Five friends go to a Cineplex which contains 6 theatres each screening a different movie and 2 other theatres screening the current blockbuster. Find the number of ways the friends can watch a movie in each of the following cases:

    (i) two of the friends must be together;

    (ii) the theatres do not matter, only the movies do.

8.4 Find the number of ways of distributing 8 distinct objects into 3 distinct boxes if each box must hold at least 2 objects.

8.5 Suppose that m distinct objects are to be distributed into n distinct boxes so that each box contains at least one object. State a restriction on m with respect to n. In how many ways can the distribution be done if

(i) $m = n$?

(ii) $m = n + 1$?

(iii) $m = n + 2$?

Chapter 9

Other Variations of the Distribution Problem

Two cases of the distribution problem were discussed in the preceding chapters. In this chapter, we shall study some of their variations.

When *identical* objects are placed in distinct boxes, whether the objects in each box are *ordered* or not makes no difference. The situation is no longer the same if the objects are distinct as shown in Figure 9.1.

Figure 9.1

In Chapter 8, we did not consider the ordering of objects in each box. In our next example, we shall take it into account.

Example 9.1 *Suppose that 5 distinct objects a, b, c, d, e are distributed into 3 distinct boxes, and that the ordering of objects in each box matters. In how many ways can this be done?*

Solution First, consider a (say). Clearly, there are 3 choices of a box for a to be put in (say, a is put in box (2)). Next, consider b. The object b can be put in one of the 3 boxes. The situation is special if b is put in box (2) because of the existence of a in that box. As the ordering of objects in each box matters, if b is put in box (2), then there are two choices for b, namely, left of a or right of a as indicated in Figure 9.2. Thus, altogether, there are 4 choices for b.

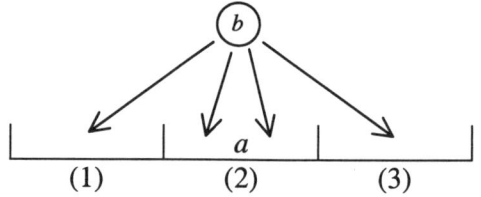

Figure 9.2

Assume that b is put in box (3). Now, consider c. As shown in Figure 9.3, c has 5 choices.

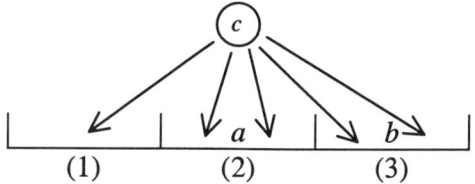

Figure 9.3

Continuing in this manner, we see that d and e have, respectively, 6 and 7 choices. Thus, the answer is given by $3 \cdot 4 \cdot 5 \cdot 6 \cdot 7$. $\qquad\square$

Let us try a different approach to solve the above problem. First, we pretend that the objects a, b, c, d, and e are all identical. The number of ways of distributing 5 identical objects into 3 distinct boxes is, by result (6.1), $\binom{5+3-1}{5}$, i.e. $\binom{7}{2}$. Next, take such a way of distribution, say,

$$\underline{|\ 0\ 0\ ||\ 0\ 0\ 0\ |}$$
$$(1)(2)(3).$$

Since the 5 objects are actually distinct and the ordering of objects matters, such a distribution for *identical* objects corresponds to 5! different distributions of *distinct* objects. Thus, by (MP), the answer is given by $\binom{7}{2} \cdot 5!$ which agrees with the answer $3 \cdot 4 \cdot 5 \cdot 6 \cdot 7$.

In general, we have:

The number of ways of distributing r *distinct* objects into n *distinct* boxes such that the ordering of objects in each box matters is given by

$$\binom{r+n-1}{r} \cdot r!$$ (9.1)

which is equal to

$$n(n+1)(n+2)\ldots(n+r-1).$$

In our previous discussion on the distribution problem, objects were either *all identical* or *all distinct*. We now consider a case that is a mixture of these two.

Example 9.2 *Four identical objects "a", three identical objects "b" and two identical objects "c" are to be distributed into 9 distinct boxes so that each box contains one object. In how many ways can this be done?*

Solution Let's start with the four a's. Among the 9 boxes, we choose 4 of them, and put one a in each chosen box. Next, we consider the three b's. From among the 5 remaining boxes, we choose 3, and put one b in each chosen box (see Figure 9.4). Finally, we put one c in each of the 2 remaining boxes.

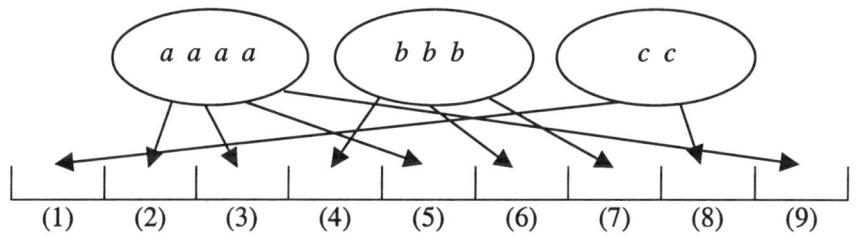

Figure 9.4

There are $\binom{9}{4}$ ways for step 1, $\binom{5}{3}$ ways for step 2 and $\binom{2}{2}(=1)$ way for step 3. Thus, by (MP), the answer is given by

$$\binom{9}{4}\binom{5}{3} \cdot 1 = \frac{9!}{4!5!} \cdot \frac{5!}{3!2!} = \frac{9!}{4!3!2!}. \qquad \square$$

Remark   In the above solution, a is considered first, followed by b and finally c. The answer is independent of this order. For instance, if b is considered first, followed by c and then a, by applying a similar argument we arrive at $\binom{9}{3}\binom{6}{2}\binom{4}{4}$, which is again $\frac{9!}{4!3!2!}$.

There is a 1–1 correspondence between the distributions considered in Example 9.2 and the arrangements of 4 a's, 3 b's and 2 c's in a row as shown in Figure 9.5.

| a | c | a | b | a | a | b | b | c | $\longleftrightarrow$ *acabaabbc*

| c | c | b | b | b | a | a | a | a | $\longleftrightarrow$ *ccbbbaaaa*

Figure 9.5

Thus, by the result of Example 9.2, the number of arrangements of 4 a's, 3 b's and 2 c's in a row is given by

$$\frac{9!}{4!3!2!}.$$

In general,

> Suppose there are n_1 identical objects of type 1, n_2 identical objects of type 2, …, and n_k identical objects of type k. Let $n = n_1 + n_2 + \ldots + n_k$. Then the number of arrangements of these n objects in a row is given by
>
> $$\binom{n}{n_1}\binom{n-n_1}{n_2}\cdots\binom{n-n_1-\cdots-n_{k-1}}{n_k}. \qquad (9.2)$$
>
> which is equal to
>
> $$\frac{n!}{n_1!n_2!\ldots n_k!}.$$

Let us reconsider Example 9.1. We observe that there is a 1–1 correspondence between the distributions considered in Example 9.1 and the arrangements of a, b, c, d, e and two 1's as shown in Figure 9.6.

By the above result, the number of arrangements of a, b, c, d, e and two 1's is given by $\frac{7!}{2!}$, which agrees also with the earlier two answers.

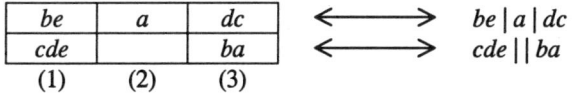

Figure 9.6

Exercise

9.1   Calculate the number of different arrangements which can be made using all the letters of the word BANANARAMA.

9.2   Calculate the number of distinct 8-letter arrangements which can be made with letters of the word INFINITE. How many of these begin with III?

(C)

9.3   Find the number of arrangements of 4 identical squares, 5 identical pentagons and 6 identical hexagons in a row if

    (i) there is no restriction;

    (ii) no two pentagons are adjacent;

    (iii) any two squares are separated by at least two other polygons.

9.4   Let $A = \{1, 2, \ldots, m\}$ and $B = \{1, 2, \ldots, n\}$ where $m, n \geq 1$. Find the number of

    (i) mappings from A to B;

    (ii) 1–1 mappings from A to B (here $m \leq n$);

    (iii) mappings $f : A \to B$ such that $f(i) < f(j)$ in B whenever $i < j$ in A (here $m \leq n$);

    (iv) mappings $f : A \to B$ such that $f(1) = 1$.

9.5   Let $A = \{1, 2, \ldots, m\}$ and $B = \{1, 2, \ldots, n\}$. Find the number of onto mappings from A to B in each of the following cases:

    (i) $m = n$;

    (ii) $m = n + 1$;

    (iii) $m = n + 2$.

(Compare this problem with Problem 8.5.)

9.6   Ten cars take part in an Automobile Association of Singapore autoventure to Malaysia. At the causeway, 4 immigration counters are open. In how many ways can the 10 cars line up in a 4-line queue?

9.7 Solve Problem 8.5 with an additional condition that the ordering of objects in each box counts.

9.8 Show that

$$\binom{n}{n_1}\binom{n-n_1}{n_2}\cdots\binom{n-n_1-\cdots-n_{k-1}}{n_k} = \frac{n!}{n_1!n_2!\cdots n_k!},$$

where $n = n_1 + n_2 + \cdots + n_k$.

Chapter 10

The Binomial Expansion

In Chapter 3, we introduced a family of numbers which were denoted by $\binom{n}{r}$ or C_r^n. Given integers n and r with $0 \leq r \leq n$, the number $\binom{n}{r}$ is defined as the number of r-element subsets of the set $N_n = \{1, 2, \ldots, n\}$. That is, $\binom{n}{r}$ is the number of ways of selecting r distinct objects from a set of n distinct objects. We also derived the following formula for $\binom{n}{r}$:

$$\binom{n}{r} = \frac{n!}{r!(n-r)!}. \qquad (10.1)$$

By applying (10.1), or otherwise, we can easily derive some interesting identities involving these numbers such as:

$$\binom{n}{r} = \binom{n}{n-r}, \qquad (10.2)$$

$$\binom{n}{r} = \binom{n-1}{r-1} + \binom{n-1}{r}, \qquad (10.3)$$

$$r\binom{n}{r} = n\binom{n-1}{r-1}, \quad r \geq 1, \qquad (10.4)$$

$$\binom{n}{m}\binom{m}{r} = \binom{n}{r}\binom{n-r}{m-r}. \qquad (10.5)$$

In this chapter, we shall learn more about this family of numbers and derive some other important identities involving them.

In algebra, we learn how to expand the algebraic expression $(1+x)^n$ for $n = 0, 1, 2, 3$. Their expansions are shown below:

$$(1+x)^0 = 1\,,$$
$$(1+x)^1 = 1+x\,,$$
$$(1+x)^2 = 1+2x+x^2\,,$$
$$(1+x)^3 = 1+3x+3x^2+x^3\,.$$

Notice that the coefficients in the above expansions are actually numbers of the form $\binom{n}{r}$. Indeed, we have:

$$1 = \binom{0}{0}\,,$$

$$1 = \binom{1}{0}\,, \quad 1 = \binom{1}{1}\,,$$

$$1 = \binom{2}{0}\,, \quad 2 = \binom{2}{1}\,, \quad 1 = \binom{2}{2}\,,$$

$$1 = \binom{3}{0}\,, \quad 3 = \binom{3}{1}\,, \quad 3 = \binom{3}{2}\,, \quad 1 = \binom{3}{3}\,.$$

What can we say about the coefficients in the expansion of $(1+x)^4$? Will we obtain

$$(1+x)^4 = \binom{4}{0} + \binom{4}{1}x + \binom{4}{2}x^2 + \binom{4}{3}x^3 + \binom{4}{4}x^4\,?$$

Let us try to find out the coefficient of x^2 in the expansion of $(1+x)^4$. We may write

$$(1+x)^4 = \underset{(1)}{(1+x)}\underset{(2)}{(1+x)}\underset{(3)}{(1+x)}\underset{(4)}{(1+x)}\,.$$

Table 10.1

(1)	(2)	(3)	(4)
x	x		
x		x	
x			x
	x	x	
	x		x
		x	x

Observe that in the expansion, each of the factors (1), (2), (3) and (4) contributes either 1 or x, and they are multiplied together to form a term. For instance, to obtain x^2 in the expansion, two of (1), (2), (3) and (4) contribute x and the remaining two contribute 1. How many ways can this be done? Table 10.1 shows all the possible ways, and the answer is 6.

Thus, there are 6 terms of x^2 and the coefficient of x^2 in the expansion of $(1 + x)^4$ is therefore 6. Indeed, to select two x's from four factors $(1 + x)$, there are $\binom{4}{2}$ ways (while the remaining two have no choice but to contribute "1"). Thus the coefficient of x^2 in the expansion of $(1+x)^4$ is $\binom{4}{2}$ which is 6. Using a similar argument, one can readily see that

$$(1 + x)^4 = \binom{4}{0} + \binom{4}{1} x + \binom{4}{2} x^2 + \binom{4}{3} x^3 + \binom{4}{4} x^4 .$$

In general, what can be said about the expansion of $(1+x)^n$, where n is any natural number?

Let us write

$$(1 + x)^n = \underset{(1)}{(1 + x)} \underset{(2)}{(1 + x)} \dots \underset{(n)}{(1 + x)} . \qquad (*)$$

To expand $(1 + x)^n$, we first select 1 or x from each of the n factors $(1 + x)$, and then multiply the n chosen 1's and x's together. The general term thus obtained is of the form x^r, where $0 \le r \le n$. What is the coefficient of x^r in the expansion of $(1 + x)^n$ if the like terms are grouped? This coefficient is the number of ways to form the term x^r in the product (*). To form a term x^r, we choose r factors $(1+x)$ from the

n factors $(1+x)$ in (*) and select x from each of the r chosen factors. Each of the remaining $n - r$ factors $(1+x)$ has no other option but to contribute 1. Clearly, the above selection can be done in $\binom{n}{r}$ ways. Thus, the coefficient of x^r in the expansion of $(1+x)^n$ is given by $\binom{n}{r}$. We thus arrive at the following result:

The Binomial Theorem (BT)

For any natural number n,

$$(1+x)^n = \binom{n}{0} + \binom{n}{1}x + \binom{n}{2}x^2 + \cdots \qquad (10.6)$$

$$+ \binom{n}{r}x^r + \cdots + \binom{n}{n}x^n.$$

Exercise

10.1   By applying Identity (10.1), or otherwise, derive the following identities:

(i) $\binom{n}{r} = \binom{n-1}{r-1} + \binom{n-1}{r}$;

(ii) $\binom{n}{m}\binom{m}{r} = \binom{n}{r}\binom{n-r}{m-r}$.

10.2   In the expansion of $(1+x)^{100}$, it is known that the coefficients of x^r and x^{3r}, where $1 \leq r \leq 33$, are equal. Find the value of r.

10.3   What is the largest value of k such that there is a binomial expansion $(1+x)^n$ in which the coefficients of k consecutive terms are in the ratio $1 : 2 : 3 : \ldots : k$? Identify the corresponding expansion and the terms.

Chapter 11

Some Useful Identities

We gave four simple identities involving binomial coefficients, namely
(10.2)–(10.5), in Chapter 10. In this chapter, we shall derive some more
identities involving binomial coefficients from (BT). These identities,
while interesting in their own right, are also useful in simplifying certain
algebraic expressions.

Consider the expansion of $(1 + x)^n$ in (BT). If we let $x = 1$, we then
obtain from (BT) the following

$$\binom{n}{0} + \binom{n}{1} + \binom{n}{2} + \cdots + \binom{n}{n} = 2^n. \qquad \text{(B1)}$$

Example 11.1 In Example 5.2, we discussed a counting problem on
$\mathcal{P}(S)$, the set of all subsets of a finite set S. If S is an n-element
set (i.e. $|S| = n$), it can be shown (see Problem 5.3) by establishing
a bijection between $\mathcal{P}(S)$ and the set of n-digit binary sequences that
there are exactly 2^n subsets of S inclusive of the empty set ϕ and the
set S itself (i.e. $|\mathcal{P}(S)| = 2^n$). We can now give a more *natural* proof
for this fact. Assume that $|S| = n$. By definition, the number of

$$0\text{-element subsets of } S \text{ is } \binom{n}{0},$$

$$1\text{-element subsets of } S \text{ is } \binom{n}{1},$$

2-element subsets of S is $\begin{pmatrix} n \\ 2 \end{pmatrix}$,

$$\vdots$$

n-element subsets of S is $\begin{pmatrix} n \\ n \end{pmatrix}$.

Thus,

$$|\mathcal{P}(S)| = \begin{pmatrix} n \\ 0 \end{pmatrix} + \begin{pmatrix} n \\ 1 \end{pmatrix} + \begin{pmatrix} n \\ 2 \end{pmatrix} + \cdots + \begin{pmatrix} n \\ n \end{pmatrix}$$

$$= 2^n \quad \text{(by (B1))}. \qquad \square$$

Example 11.2 *The number 4 can be expressed as a sum of one or more positive integers, taking order into account, in the following 8 ways:*

$$4 = 4 = 1 + 3$$
$$= 3 + 1 = 2 + 2$$
$$= 1 + 1 + 2 = 1 + 2 + 1$$
$$= 2 + 1 + 1 = 1 + 1 + 1 + 1.$$

Show that every natural number n can be so expressed in 2^{n-1} ways.

Solution This is in fact Problem 5.5. Let us see how (B1) can be used to prove the result. But first of all, consider the special case above when $n = 4$.

We write $4 = 1 + 1 + 1 + 1$ and note that there are three "+"s in the expression. Look at the following relation.

$$4 \longleftrightarrow \underbrace{1 + 1 + 1 + 1}_{4} \qquad \text{(no "+" is chosen)}$$

$$1 + 3 \longleftrightarrow \underbrace{1}_{1} \oplus \underbrace{1 + 1 + 1}_{3} \left.\begin{array}{l} \\ \\ \\ \\ \\ \end{array}\right\}$$

$$3 + 1 \longleftrightarrow \underbrace{1 + 1 + 1}_{3} \oplus \underbrace{1}_{1} \quad \text{(one "+" is chosen)}$$

$$2 + 2 \longleftrightarrow \underbrace{1 + 1}_{2} \oplus \underbrace{1 + 1}_{2}$$

$$1+1+2 \longleftrightarrow \underbrace{1}_{1} \oplus \underbrace{1}_{1} \oplus \underbrace{1+1}_{2}$$

$$1+2+1 \longleftrightarrow \underbrace{1}_{1} \oplus \underbrace{1+1}_{2} \oplus \underbrace{1}_{1}$$ (2 "+"s are chosen)

$$2+1+1 \longleftrightarrow \underbrace{1+1}_{2} \oplus \underbrace{1}_{1} \oplus \underbrace{1}_{1}$$

$$1+1+1+1 \longleftrightarrow \underbrace{1}_{1} \oplus \underbrace{1}_{1} \oplus \underbrace{1}_{1} \oplus \underbrace{1}_{1}$$ (3 "+"s are chosen)

This correspondence is actually a bijection between the set of all such expressions of 4 and the set of all subsets of three "+"s. Thus, by (BP) and (B1), the required answer is

$$\binom{3}{0} + \binom{3}{1} + \binom{3}{2} + \binom{3}{3} = 2^3 .$$

In general, write

$$n = \underbrace{1 + 1 + \cdots + 1 + 1}_{n}$$

and note that there are $n-1$ "+"s in the above expression. We now extend the above technique by establishing a bijection between the set of all such expressions of n and the set of all subsets of $n-1$ "+"s. Thus, by (BP) and (B1), the number of all such expressions of n is

$$\binom{n-1}{0} + \binom{n-1}{1} + \cdots + \binom{n-1}{n-1} = 2^{n-1} . \qquad \square$$

Consider again the expansion of $(1+x)^n$ in (BT). If we now let $x = -1$, we then have

$$\binom{n}{0} - \binom{n}{1} + \binom{n}{2} - \binom{n}{3} \cdots + (-1)^n \binom{n}{n} = 0 ,$$

where the terms on the LHS alternate in sign. Thus, if n is even, say $n = 2k$, then

$$\binom{n}{0} + \binom{n}{2} + \cdots + \binom{n}{2k} = \binom{n}{1} + \binom{n}{3} + \cdots + \binom{n}{2k-1} ;$$

and if n is odd, say $n = 2k + 1$, then

$$\binom{n}{0} + \binom{n}{2} + \cdots + \binom{n}{2k} = \binom{n}{1} + \binom{n}{3} + \cdots + \binom{n}{2k+1}.$$

As

$$\left[\binom{n}{0} + \binom{n}{2} + \cdots\right] + \left[\binom{n}{1} + \binom{n}{3} + \cdots\right] = 2^n$$

by (B1), we have:

$$\binom{n}{0} + \binom{n}{2} + \binom{n}{4} + \cdots = \binom{n}{1} + \binom{n}{3} + \binom{n}{5} + \cdots$$

$$= \frac{1}{2}(2^n) = 2^{n-1}. \tag{B2}$$

Example 11.3 *A finite set S is said to be "even" ("odd") if $|S|$ is even (odd). Consider $N_8 = \{1, 2, \ldots, 8\}$. How many even (odd) subsets does N_8 have?*

Solution The number of even subsets of N_8 is

$$\binom{8}{0} + \binom{8}{2} + \binom{8}{4} + \binom{8}{6} + \binom{8}{8},$$

and the number of odd subsets of N_8 is

$$\binom{8}{1} + \binom{8}{3} + \binom{8}{5} + \binom{8}{7}.$$

By (B2),

$$\binom{8}{0} + \binom{8}{2} + \cdots + \binom{8}{8} = \binom{8}{1} + \binom{8}{3} + \binom{8}{5} + \binom{8}{7}$$

$$= 2^{8-1} = 2^7 = 128. \qquad \square$$

Consider the following binomial expansion once again:

$$(1 + x)^n = \binom{n}{0} + \binom{n}{1}x + \binom{n}{2}x^2 + \binom{n}{3}x^3 + \cdots + \binom{n}{n}x^n.$$

If we treat the expressions on both sides as functions of x, and differentiate them with respect to x, we obtain:

$$n(1+x)^{n-1} = \binom{n}{1} + 2\binom{n}{2}x + 3\binom{n}{3}x^2 + \cdots + n\binom{n}{n}x^{n-1}.$$

By letting $x = 1$ in the above identity, we have:

$$\sum_{k=1}^{n} k\binom{n}{k} = \binom{n}{1} + 2\binom{n}{2} + 3\binom{n}{3} + \cdots + n\binom{n}{n}$$
$$= n2^{n-1}.$$

(B3)

Let us try to derive (B3) by a different way. Consider the following problem. Suppose that there are $n(n \geq 1)$ people in a group, and they wish to form a committee consisting of people from the group, including the selection of a leader for the committee. In how many ways can this be done?

Let us illustrate the case when $n = 3$. Suppose that A, B, C are the three people in the group, and that a committee consists of k members from the group, where $1 \leq k \leq 3$. For $k = 1$, there are 3 ways to do so as shown below.

Committee members	Leader
A	A
B	B
C	C

For $k = 2$, there are 6 ways to do so as shown below.

Committee members	Leader
A, B	A
A, B	B
A, C	A
A, C	C
B, C	B
B, C	C

For $k = 3$, there are 3 ways to do so as shown below.

Committee members	Leader
A, B, C	A
A, B, C	B
A, B, C	C

Thus, there are altogether $3 + 6 + 3 = 12$ ways to do so.

In general, from a group of n people, there are $\binom{n}{k}$ ways to form a k-member committee, and k ways to select a leader from the k members in the committee. Thus, the number of ways to form a k-member committee including the selection of a leader is, by (MP), $k\binom{n}{k}$. As k could be $1, 2, \ldots, n$, by (AP), the number of ways to do so is given by

$$\sum_{k=1}^{n} k \binom{n}{k}.$$

Let us count the same problem via a different approach as follows. First, we select a leader from the group, and then choose $k-1$ members, where $k = 1, 2, \ldots, n$, from the group to form a k-member committee. There are n choices for the first step and

$$\binom{n-1}{0} + \binom{n-1}{1} + \cdots + \binom{n-1}{n-1}$$

ways for the second step. Thus, by (MP) and (B1), the required number is

$$n\left[\binom{n-1}{0} + \binom{n-1}{1} + \cdots + \binom{n-1}{n-1}\right] = n2^{n-1}.$$

Since both

$$\sum_{k=1}^{n} k \binom{n}{k} \quad \text{and} \quad n2^{n-1}$$

count the same number, identity (B3) follows.

In the above discussion, we establish identity (B3) by first introducing a "suitable" counting problem. We then count the problem in two different ways so as to obtain two different expressions. These two different expressions must be equal as they count the same quantity. This way of deriving an identity is quite a common practice in combinatorics, and is known as "counting it twice".

Exercise

11.1 By applying Identity (10.5) or otherwise, show that

$$\sum_{k=r}^{n} \binom{n}{k}\binom{k}{r} = \binom{n}{r} 2^{n-r}, \quad \text{where } 0 \le r \le n.$$

11.2 Show that

$$\sum_{k=0}^{n-1} \binom{2n-1}{k} = 2^{2n-2}.$$

11.3 Show that

$$\sum_{k=0}^{n} \frac{1}{k+1}\binom{n}{k} = \frac{1}{n+k}(2^{n+1}-1)$$

by integrating both sides of $(1+x)^n = \sum_{k=0}^{n} \binom{n}{k}x^k$ with respect to x.

11.4 Show that

$$\sum_{k=1}^{n} k^2 \binom{n}{k} = n(n+1)2^{n-2}.$$

11.5 Solve Example 11.2 by using result (7.1)(ii).

Chapter 12

Pascal's Triangle

In Chapter 10, we established the Binomial Theorem (BT) which states that for all nonnegative integers n,

$$(1+x)^n = \sum_{r=0}^{n} \binom{n}{r} x^r.$$

Let us display the binomial coefficients row by row following the increasing values of n as shown in Figure 12.1. We observe from Figure 12.1 the following.

1. The binomial coefficient at a lattice point counts the number of shortest routes from the top lattice point (representing $\binom{0}{0}$) to the lattice point concerned. For example, there are $\binom{4}{2}$ $(= 6)$ shortest

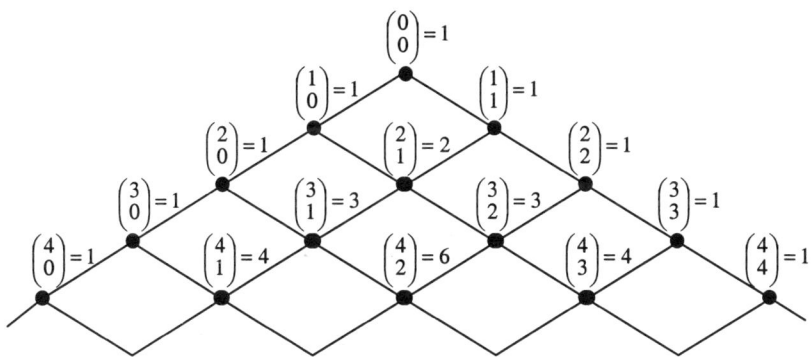

Figure 12.1

routes from the lattice point representing $\binom{0}{0}$ to the lattice point $\binom{4}{2}$ (also see Example 5.1).

2. The number pattern is symmetric with respect to the vertical line through the top lattice point, and this observation corresponds to the identity $\binom{n}{r} = \binom{n}{n-r}$ (see (10.2)).

3. Any binomial coefficient represented by an interior lattice point is equal to the sum of the two binomial coefficients represented by the lattice points on its "shoulders" (see Figure 12.2). This observation corresponds to the identity $\binom{n}{r} = \binom{n-1}{r-1} + \binom{n-1}{r}$ (see (10.3)).

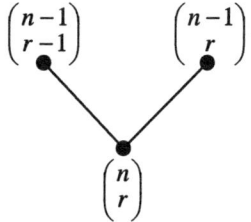

Figure 12.2

4. The sum of the binomial coefficients in the nth row is equal to 2^n and this fact corresponds to the identity

$$\binom{n}{0} + \binom{n}{1} + \binom{n}{2} + \cdots + \binom{n}{n} = 2^n.$$

The number pattern of Figure 12.1 was known to Omar Khayyam and Jia Xian around 1100 AD. The pattern was also found in the book written by the Chinese mathematician Yang Hui in 1261, in which Yang Hui called it, the Jia Xian triangle. The number pattern in the form of Figure 12.3 was found in another book written by another Chinese mathematician Zhu Shijie in 1303.

However, the number pattern of Figure 12.1 is generally called *Pascal's Triangle* in memory of the great French mathematician Blaise Pascal (1623–1662) who also applied the "triangle" to the study of *probability*, a subject dealing with "chance". For a history of this number pattern, readers are referred to the book *Pascal's Arithmetical Triangle* by A. W. F. Edwards (Oxford University Press (1987)).

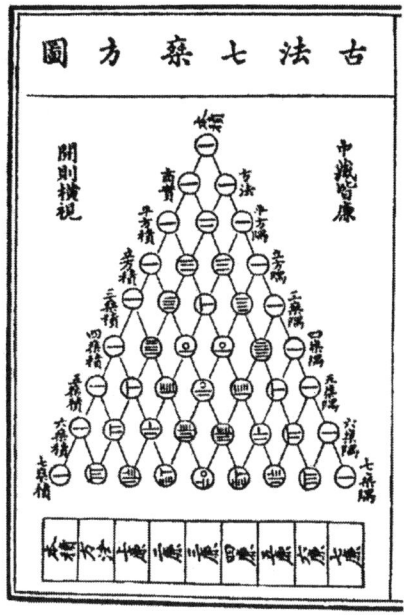

Figure 12.3

Blaise Pascal

Look at Pascal's triangle of Figure 12.4.

What is the sum of the six binomial coefficients enclosed in the shaded rectangle? The answer is 56. Note that this answer appears as another binomial coefficient located at the right side of 21 in the next row. Is this situation just a coincidence? Let us take a closer look.

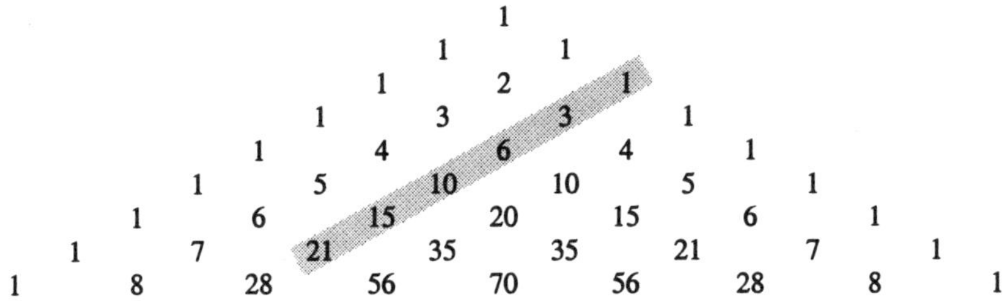

Figure 12.4

Observe that

$$1 + 3 + 6 + 10 + 15 + 21 = \binom{2}{2} + \binom{3}{2} + \binom{4}{2} + \binom{5}{2} + \binom{6}{2} + \binom{7}{2}$$

$$= \binom{3}{3} + \binom{3}{2} + \binom{4}{2} + \binom{5}{2} + \binom{6}{2} + \binom{7}{2}$$

$$\left(\text{as } \binom{2}{2} = \binom{3}{3} \right)$$

$$= \binom{4}{3} + \binom{4}{2} + \binom{5}{2} + \binom{6}{2} + \binom{7}{2}$$

$$= \binom{5}{3} + \binom{5}{2} + \binom{6}{2} + \binom{7}{2}$$

$$= \binom{6}{3} + \binom{6}{2} + \binom{7}{2}$$

$$= \binom{7}{3} + \binom{7}{2}$$

$$= \binom{8}{3} \quad (= 56)$$

by applying the identity $\binom{n-1}{r-1} + \binom{n-1}{r} = \binom{n}{r}$.

The above result is really a special case of a general situation. As a matter of fact, the above argument can also be used to establish the following general result (also see Figure 12.5):

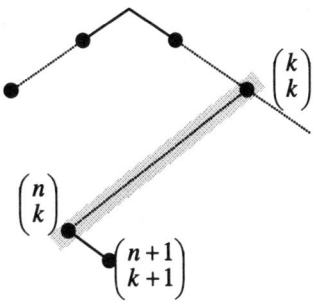

Figure 12.5

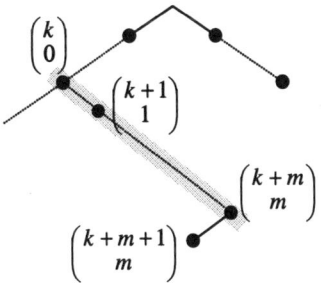

Figure 12.6

For any nonnegative integers n and k with $n \geq k$,

$$\binom{k}{k} + \binom{k+1}{k} + \cdots + \binom{n}{k} = \binom{n+1}{k+1}.$$

(B4)

By the symmetry of Pascal's triangle, one obtains the following accompanying identity of (B4) (also see Figure 12.6):

$$\binom{k}{0} + \binom{k+1}{1} + \cdots + \binom{k+m}{m} = \binom{k+m+1}{m}.$$

(B5)

To end this chapter, we show an application of identity (B4) in the solution of the following problem which appeared in International Mathematical Olympiad 1981.

Example 12.1   *Let $1 \leq r \leq n$ and consider all r-element subsets of the set $\{1, 2, \ldots, n\}$. Each of these subsets has a smallest member. Let $F(n, r)$ denote the arithmetic mean of these smallest numbers. Prove that*

$$F(n, r) = \frac{n + 1}{r + 1}.$$

Solution   As an illustration of this problem, we consider the case when $n = 6$ and $r = 4$. There are $\binom{6}{4} (= 15)$ 4-element subsets of the set $\{1, 2, 3, 4, 5, 6\}$. They and their "smallest members" are listed in Table 12.1.

Table 12.1

4-element subsets of $\{1, 2, \ldots, 6\}$	Smallest member
$\{1, 2, 3, 4\}$	1
$\{1, 2, 3, 5\}$	1
$\{1, 2, 3, 6\}$	1
$\{1, 2, 4, 5\}$	1
$\{1, 2, 4, 6\}$	1
$\{1, 2, 5, 6\}$	1
$\{1, 3, 4, 5\}$	1
$\{1, 3, 4, 6\}$	1
$\{1, 3, 5, 6\}$	1
$\{1, 4, 5, 6\}$	1
$\{2, 3, 4, 5\}$	2
$\{2, 3, 4, 6\}$	2
$\{2, 3, 5, 6\}$	2
$\{2, 4, 5, 6\}$	2
$\{3, 4, 5, 6\}$	3

By definition,

$$F(6, 4) = (10 \cdot 1 + 4 \cdot 2 + 1 \cdot 3) \div 15$$

$$= \frac{7}{5},$$

and this is equal to $\frac{n+1}{r+1}$ when $n = 6$ and $r = 4$.

Write $N_n = \{1, 2, \ldots, n\}$. To evaluate $F(n, r)$, it is clear that we need to first find out

1. which numbers in N_n could be the smallest member of an r-element subset of N_n (in the above example, these are 1, 2, 3 but not 4, 5, 6), and
2. how many times such a smallest member occurs (in the above example, 1 occurs ten times, 2 four times and 3 once);

and then sum these smallest numbers up, and finally divide by $\binom{n}{r}$, the number of r-element subsets of N_n, to obtain the "average".

The last r elements (according to the magnitude) of the set N_n are:

$$\underbrace{n-r+1, n-r+2, \ldots, n-r+r (=n)}_{r} .$$

It follows that $n - r + 1$ is the *largest* possible number to be the smallest member of an r-element subset of N_n. Hence, $1, 2, 3, \ldots, n - r + 1$ are all the possible candidates to be the smallest members of r-element students of N_n.

Let $k \in \{1, 2, 3, \ldots, n-r+1\}$. Our next task is to find out how many times k occurs as the smallest member. To form an r-element subset of N_n containing k as the smallest member, we simply form an $(r-1)$-element subset from the $(n-k)$-element set $\{k+1, k+2, \ldots, n\}$ and then add k to it. The number of $(r-1)$-element subsets of $\{k+1, k+2, \ldots, n\}$ is given by $\binom{n-k}{r-1}$. Thus, k occurs $\binom{n-k}{r-1}$ times as the smallest member. Let Σ denote the sum of all these smallest members. Then, as $k = 1, 2, \ldots, n - r + 1$, we have

$$\Sigma = 1\binom{n-1}{r-1} + 2\binom{n-2}{r-1} + 3\binom{n-3}{r-1} + \cdots + (n-r+1)\binom{n-(n-r+1)}{r-1}$$

$$= (n-r+1)\binom{r-1}{r-1} + \cdots + 3\binom{n-3}{r-1} + 2\binom{n-2}{r-1} + 1\binom{n-1}{r-1}$$

$$\left.\begin{aligned} &= \binom{r-1}{r-1} + \cdots + \binom{n-3}{r-1} + \binom{n-2}{r-1} + \binom{n-1}{r-1} \\ &\quad + \binom{r-1}{r-1} + \cdots + \binom{n-3}{r-1} + \binom{n-2}{r-1} \\ &\quad + \binom{r-1}{r-1} + \cdots + \binom{n-3}{r-1} \\ &\qquad \vdots \\ &\quad + \binom{r-1}{r-1} . \end{aligned}\right\} \begin{array}{l} n-r+1 \text{ rows of} \\ \text{summands} \end{array}$$

Now, by applying (B4) to each summand above except the last one and noting that $\binom{r-1}{r-1} = \binom{r}{r}$, Σ can be simplified to

$$\Sigma = \underbrace{\binom{n}{r} + \binom{n-1}{r} + \cdots + \binom{r}{r}}_{n-r+1}.$$

By applying (B4) once again, we have

$$\Sigma = \binom{n+1}{r+1}.$$

Finally, by definition of $F(n, r)$, it follows that

$$F(n, r) = \Sigma \div \binom{n}{r} = \binom{n+1}{r+1} \div \binom{n}{r}$$

$$= \frac{(n+1)!}{(r+1)!(n-r)!} \cdot \frac{r!(n-r)!}{n!}$$

$$= \frac{n+1}{r+1}$$

as desired.   □

Exercise

12.1   Find the coefficient of x^5 in the expansion of

$$(1+x)^5 + (1+x)^6 + \cdots + (1+x)^{100}.$$

12.2   Consider the rows of Pascal's Triangle. Prove that if a row is made into a single number by using each element as a digit of the number (carrying over when an element itself has more than one digit), the number is equal to 11^{n-1}. (For example, from the first row $1 = 11^0$, from the second row $11 = 11^1$, from the third row $121 = 11^2$, and from the 6th row $15(10)(10)51 = 15(11)051 = 161051 = 11^5$.)

12.3   On the rth day of an army recruitment exercise, r men register themselves. Each day, the recruitment officer chooses exactly

k of the men and line them up in a row to be marched to the barracks. Show that the sum of the numbers of all the possible rows in the first $2k$ days is equal to the number of possible rows in the $(2k + 1)$th day.

12.4 The greatest integer not exceeding a real number x is denoted by $\lfloor x \rfloor$. Show that

(i) $\binom{n}{i} < \binom{n}{j}$ if $0 \le i < j \le \lfloor \frac{n}{2} \rfloor$;

(ii) $\binom{n}{i} \ge \binom{n}{j}$ if $\lfloor \frac{n}{2} \rfloor \le i < j \le n$, with equality if and only if $i = \lfloor \frac{n}{2} \rfloor, j = \lfloor \frac{n}{2} \rfloor + 1$ and n is odd.

12.5 Evaluate $n! + \frac{(n+1)!}{1!} + \frac{(n+2)!}{2!} + \cdots + \frac{(n+r)!}{r!} + \cdots + \frac{(3n)!}{(2n)!}$.

Chapter 13

Miscellaneous Problems

13.1 One commercially available ten-button lock may be opened by depressing — in any order — the correct five buttons. The sample shown below has $\{1, 2, 3, 6, 9\}$ as its combination. Suppose that these locks are redesigned so that sets of as many as nine buttons or as few as one button could serve as combinations. How many additional combinations would this allow?

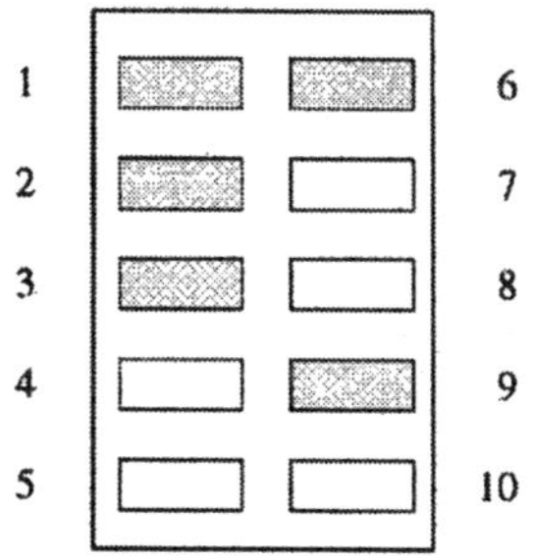

(AIME)

13.2 Calculate in how many ways each of the following choices can be made.

(i) 4 books are to be chosen from a list of 10 titles to be taken away for reading during a holiday.

(ii) 20 people have sent in winning entries for a newspaper competition, and three are to be chosen and placed in order of merit so as to receive the 1st, 2nd and 3rd prizes.

(iii) A team of 6 people is to be chosen from a list of 10 possibles; the team consists of a 1st pair, a 2nd pair and a 3rd pair, but order within each pair does not matter.

(C)

13.3   A society is planning a ballot for the office of president. There are 5 candidates for the office. In order to eliminate the order of the candidates on the ballot as a possible influence on the election, there is a rule that on the ballot slips, each candidate must appear in each position the same number of times as any other candidate. What is the smallest number of different ballot slips necessary?

13.4   In the waiting area of a specialist clinic, patients sit on chairs arranged 10 to a row with an aisle on either side. Ten patients are sitting in the second row. How many ways are there for all the patients in the second row to see the doctor if at least one patient has to pass over one or more other patients in order to reach an aisle?

13.5   In how many ways can 4 a's, 4 b's, 4 c's and 4 d's be arranged in a 4×4 array so that exactly one letter occurs in each row and in each column? (Such an arrangement is called a Latin square.)

13.6   A card is drawn from a full pack of 52 playing cards. If the card is a King, Queen or Jack, two dice are thrown and the total T is taken to be the sum of the scores on the dice. If any other card is drawn, only one die is thrown and T is taken to be the sum of the scores on the card (an Ace is considered as 1) and the die. Find the number of ways for each of the following:

(i) $T \leq 2$;
(ii) $T \geq 13$;
(iii) T is odd.

13.7   In each of the following 5-digit numbers

$$25225, 33333, 70007, 11888, \ldots$$

every digit appears more than once. Find the number of such 5-digit numbers.

13.8   The following list contains some permutations of N_9 in which each of the digits 2, 3, 4 appears in between 1 and 9:

$$814\underline{7}3\underline{6}2\underline{5}9, 569\underline{324}178, 79\underline{3}5\underline{48}2\underline{16}, \ldots .$$

Find the number of such permutations of N_9.

13.9   The following list contains some permutations of N_9 in which each of the digits 1, 2, 3 appears to the right of 9:

$$45897\underline{12}6\underline{3}, 6954\underline{38}1\underline{72}, 854796\underline{123}, \ldots .$$

Find the number of such permutations of N_9.

13.10   Find the number of 0's at the end of $1 \times 2 \times 3 \times \cdots \times 2002$.

13.11   Find the number of 15-digit ternary sequences (formed by 0, 1 and 2) in each of the following cases:

   (i)   there is no restriction;
   (ii) there are exactly three 0's;
   (iii) there are exactly four 0's and five 1's;
   (iv) there are at most two 0's;
   (v)   there is at least one pair of consecutive digits that are the same;
   (vi) there are exactly one "00", three "11", three "22", three "02", two "21" and two "10" (for instance, 002211102221102).

13.12   Find the number of (i) positive divisors (ii) even positive divisors of 2160.

13.13   Find all natural numbers which are divisible by 30 and have exactly 30 different divisors.

13.14   Consider the following grid:

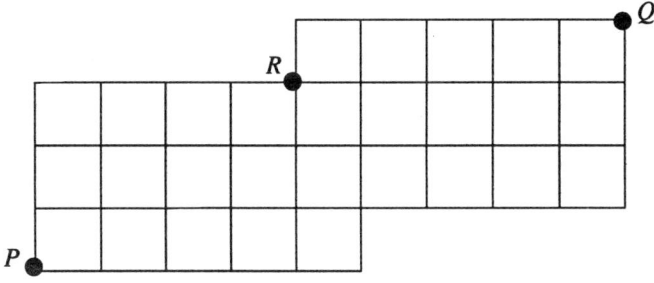

Find in the grid

(i) the number of shortest *P–R* routes;
(ii) the number of shortest *P–Q* routes.

13.15   The following figure shows 10 distinct points on the circumference of a circle.

(i) How many chords of the circle formed by these points are there?
(ii) If no three chords are concurrent within the circle, how many points of intersection of these chords within the circle are there?

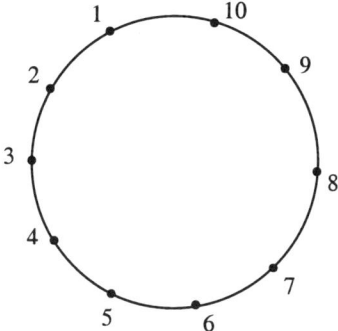

13.16   In a shooting match, eight clay targets are arranged in two hanging columns of three each and one column of two, as pictured.

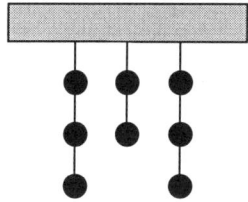

A marksman is to break all eight targets according to the following rules:

(1) The marksman first chooses a column for which a target is to be broken.

(2) The marksman must break the lowest remaining unbroken target in the chosen column.

If these rules are followed, in how many different orders can the eight targets be broken?

(AIME)

13.17 Six scientists are working on a secret project. They wish to lock up the documents in a cabinet such that the cabinet can be opened when and only when three or more of the scientists are present. What is the smallest number of locks needed? What is the smallest number of keys each scientist must carry?

13.18 A team for a boxing competition consists of a heavyweight, a middleweight and a lightweight. There are 5 teams in the competition.

(i) If each person fights with each person of a similar weight class, how many fights take place?

(ii) At the end of the competition, everyone shakes hands exactly once with every other person, except his teammates (they have to tend to each other's wounds later). How many handshakes take place?

13.19 Find the number of paths in the array which spell out the word COUNTING.

```
                            C
                        C   O   C
                    C   O   U   O   C
                C   O   U   N   U   O   C
            C   O   U   N   T   N   U   O   C
        C   O   U   N   T   I   T   N   U   O   C
    C   O   U   N   T   I   N   I   T   N   U   O   C
C   O   U   N   T   I   N   G   N   I   T   N   U   O   C
    C   O   U   N   T   I   N   I   T   N   U   O   C
        C   O   U   N   T   I   T   N   U   O   C
            C   O   U   N   T   N   U   O   C
                C   O   U   N   U   O   C
                    C   O   U   O   C
                        C   O   C
                            C
```

13.20 Let $A = \{1, 2, \ldots, 500\}$. Find

(i) the number of 2-element subsets of A;

(ii) the number of 2-element subsets $\{a, b\}$ of A such that $a \cdot b$ is a multiple of 3;

(iii) the number of 2-element subsets $\{a, b\}$ of A such that $a + b$ is a multiple of 3.

13.21   There are 7 ways to divide 4 distinct objects a, b, c, d into two nonempty groups as shown below:

$$\{a, b, c\} \cup \{d\}, \{a, b, d\} \cup \{c\}, \{a, c, d\} \cup \{b\}, \{b, c, d\} \cup \{a\},$$

$$\{a, b\} \cup \{c, d\}, \{a, c\} \cup \{b, d\}, \{a, d\} \cup \{b, c\}.$$

How many ways are there to divide n distinct objects, where $n \geq 2$, into two nonempty groups?

13.22   Two integers p and q, with $p \geq 2$ and $q \geq 2$, are said to be *coprime* if p and q have no common prime factor. Thus 8 and 9 are coprime while 4 and 6 are not.

(i) Find the number of ways to express 360 as a product of two coprime numbers (the order of these two numbers is unimportant).

(ii) In general, given an integer $n \geq 2$, how do you find the number of ways to express n as a product of two coprime numbers where the order is immaterial?

13.23   The lattice points of the following $m \times n (n \leq m)$ grid are named as shown:

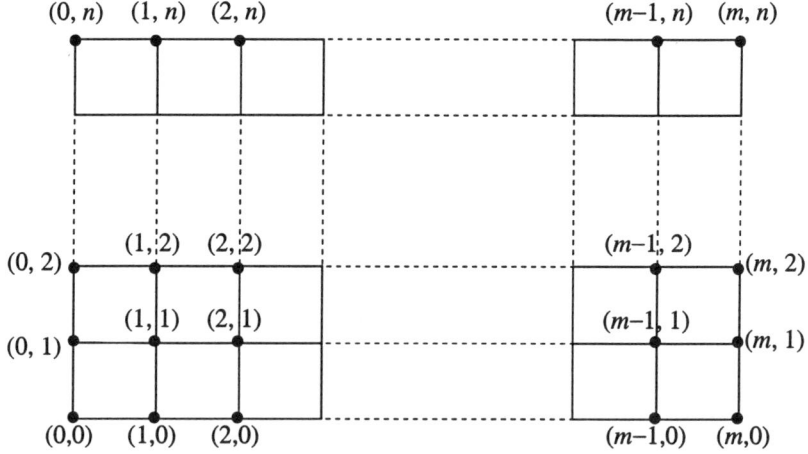

For $k \in \{1, 2, \ldots, n\}$, let p be the number of shortest $(k, k-1)$–(m, n) routes and q be the number of shortest $(k-1, k)$–(m, n) routes. Show that $p(n+1-k) = q(m+1-k)$.

13.24   The face cards (Kings, Queens and Jacks) are removed from a pack of playing cards. Six cards are drawn one at a time from this pack of cards such that they are in increasing order of magnitude. How many ways are there to do this?

13.25   There are 12 coins on a table. I pick up a number (non-zero) of coins each time. Find the number of ways of picking up all the 12 coins in the following cases:

(i) I pick up all the 12 coins in an even number of picks.
(ii) I pick up an even number of coins each time.

13.26   Find the number of 4-tuples of integers

(i) (a, b, c, d) satisfying $1 \le a < b < c < d \le 30$;
(ii) (p, q, r, s) satisfying $1 \le p \le q \le r \le s \le 30$.

13.27   Consider the following two 15-digit ternary sequences (formed by 0, 1 and 2):

$$0\ 0\ 0\ 1\ 1\ 1\ 2\ 2\ 0\ 0\ 1\ 1\ 2\ 2\ 2$$
$$0\ 1\ 2\ 2\ 2\ 0\ 0\ 0\ 0\ 1\ 1\ 1\ 1\ 2\ 2$$

Observe that each of the sequences contains exactly three 00, three 11, three 22, two 01, two 12 and one 20. Find the number of such ternary sequences.

13.28   There are n upright cups in a row. At each step, I turn over $n-1$ of them. Show that I can end up with all the cups upside down if and only if n is even. Find the number of ways this can be done in a minimum number of steps.

13.29   The following diagram shows 15 distinct points: w_1, w_2, w_3, $x_1, \ldots, x_4, y_1, \ldots, y_6, z_1, z_2$ chosen from the sides of rectangle $ABCD$.

(i) How many line segments are there joining any two points each on different sides?
(ii) How many triangles can be formed from these points?
(iii) How many quadrilaterals can be formed from these points?

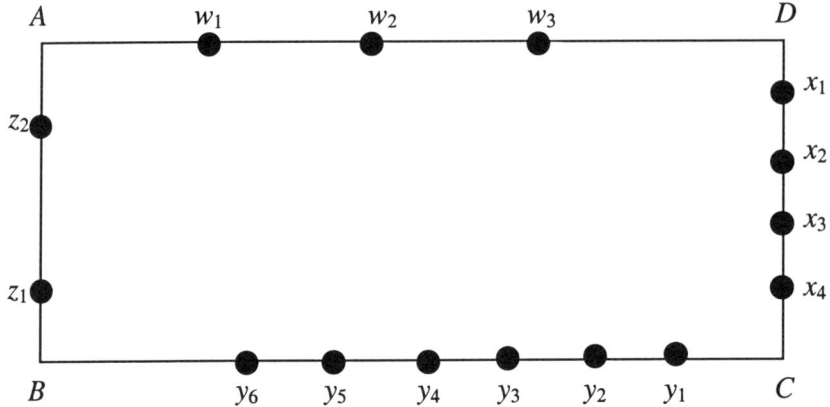

(iv) If no three line segments are concurrent in the interior of the rectangle, find the number of points of intersection of these line segments in the interior of rectangle $ABCD$.

13.30   A ternary sequence is a sequence formed by 0, 1 and 2. Let n be a positive integer. Find the number of n-digit ternary sequences

   (i) which contain at least one "0";
  (ii) which contain one "0" and one "1";
 (iii) which contain three 2's.

13.31   Each of the following six configurations consists of 4 vertices w, x, y, z with some pairs of vertices joined by lines. We are now given five colours 1, 2, 3, 4, 5 to colour the 4 vertices such that

(1) each vertex is coloured by one colour and

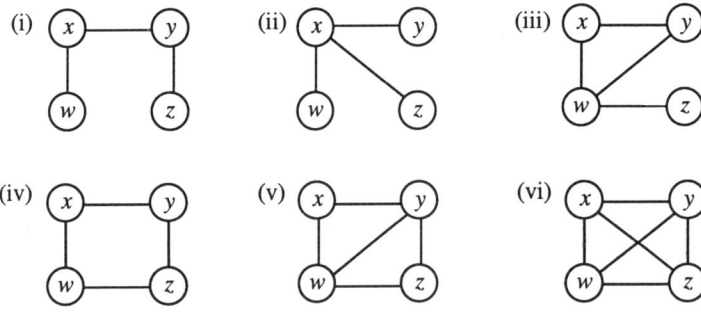

(2) any two vertices which are joined by a line must be coloured by different colours.

How many different ways are there to colour each configuration?

13.32   If repetitions are not allowed, find the number of different 5-digit numbers which can be formed from $0, 1, 2, \ldots, 9$ and are

   (i) divisible by 25;
  (ii) odd and divisible by 25;
 (iii) even and divisible by 25;
  (iv) greater than 75000;
   (v) less than 75000;
  (vi) in the interval [30000, 75000] and divisible by 5.

13.33   There are 12 boys and 8 girls, including a particular boy B and two particular girls G_1 and G_2, in a class. A class debating team of 4 speakers and a reserve is to be formed for the inter-class games. Find the number of ways this can be done if the team is to contain

    (i) exactly one girl;
   (ii) exactly two girls;
  (iii) at least one girl;
  (iv) at most two girls;
   (v) G_1;
  (vi) no B;
 (vii) B and G_1;
(viii) neither B nor G_1;
  (ix) exactly one from G_1 and G_2;
   (x) an odd number of girls.

13.34   A group of 6 people is to be chosen from 7 couples. Find the number of ways this can be done if the group is to contain

   (i) three couples;
  (ii) no couples;
 (iii) exactly one couple;
 (iv) exactly two couples;
  (v) at least one couple.

13.35   Find the number of ways in which 6 people can be divided into

(i) 3 groups consisting of 3, 2, and 1 persons;

(ii) 3 groups with 2 persons in each group;

(iii) 4 groups consisting of 2, 2, 1 and 1 persons;

(iv) 3 groups with 2 persons in each group, and the groups are put in 3 distinct rooms.

Books Recommended for Further Reading

1. K. P. Bogart, *Introductory Combinatorics* (3rd ed.), S. I. Harcourt Brace College Publishers, 1998.
2. R. A. Brualdi, *Introductory Combinatorics* (3rd ed.), Prentice Hall, 1999.
3. C. C. Chen and K. M. Koh, *Principles and Techniques in Combinatorics*, World Scientific, 1992.
4. D. I. A. Cohen, *Basic Techniques of Combinatorial Theory*, John Wiley & Sons, 1978.
5. R. L. Graham, D. E. Knuth and O. Patashnik, *Concrete Mathematics* (2nd ed.), Addison-Wesley, 1994.
6. B. W. Jackson and T. Dmitri, *Applied Combinatorics with Problem Solving*, Addison-Wesley, 1990.
7. C. L. Liu, *Introduction to Combinatorial Mathematics*, McGraw-Hill, 1968.
8. F. Roberts and B. Tesman, *Applied Combinatorics* (2nd ed.), Prentice Hall, 2002.
9. A. Tucker, *Applied Combinatorics* (4th ed.), John Wiley & Sons, 2002.

Answers to Exercises

1.1	6	**1.2**	$1, 5, 14, 55, \sum_{r=1}^{n} r^2$		
1.3	$6n - 4$	**1.4**	29	**1.5**	27
1.6	60	**1.7**	31	**1.8**	29
1.9	14	**2.1**	90	**2.2(i)**	6
2.2(ii)	15	**2.3(i)**	20	**2.3(ii)**	mn
2.3(iii)	mnt	**2.4**	30	**2.5(i)**	3^n
2.5(ii)	2^n	**2.5(iii)**	$2^{n-1}(1+n)$	**2.5(iv)**	$n^2 + n + 1$
2.6(i)	47	**2.6(ii)**	205	**2.6(iii)**	378
3.4	$(n+1)! - 1$	**4.1(v)**	$2 \cdot 9!$	**4.1(vi)**	$2 \cdot 8!$

4.1(vii) $9!3!$ **4.1(viii)** $7!5!$ **4.1(ix)** $2 \cdot 5!6!$

4.1(x) $8!9 \cdot 8 \cdot 7$ **4.1(xi)** $5!4!2$

4.1(xii) $2 \cdot 4 \cdot 5 \cdot 6 \cdot 7 \cdot 8 \cdot 9 \cdot 10 \cdot 11$ **4.2(iii)** 1008

4.2(iv) 392 **4.2(v)** 1823 **4.3** $9 \cdot 7 \cdot 5 \cdot 3$

4.4(i) $\binom{n}{6}$ **4.4(ii)** $5\binom{n}{5}$ **4.4(iii)** $4\binom{n}{4}$

4.4(iv) $\binom{n}{6} + 5\binom{n}{5} + 4\binom{n}{4} + \binom{n}{3}$ **4.5(i)** $10!$ **4.5(ii)** $8!3!$

4.5(iii) $7! \cdot 8 \cdot 7 \cdot 6$ **4.5(iv)** $7!3!70$ **4.6** $7 \cdot 5 \cdot 3$

4.7(i) $5!$ **4.7(ii)** $4!2!$ **4.7(iii)** $2!\binom{4}{2}3!$

4.8(i) $\binom{4}{2}\binom{7}{3}$ **4.8(ii)** $\binom{11}{5} - \left(\binom{7}{5} + \binom{7}{4}\binom{4}{1} + \binom{7}{3}\binom{4}{2}\right)$

4.8(iii) $\binom{11}{5} - \binom{9}{3}$ **4.9** $9 \cdot 10^{\lfloor \frac{n-1}{2} \rfloor}$ **4.10(i)** $\binom{10}{6}$

4.10(ii) P_6^{10} **4.10(iii)** $\binom{10}{2}\binom{8}{2}\binom{6}{2}$ **4.11** 162

4.12(i) $\binom{10}{7}$ **4.12(ii)** $\binom{10}{4}\binom{6}{3}\binom{3}{2}$ **4.13(i)** 405

4.13(ii) 90 **4.14** $\binom{6}{2}$ **4.14(i)** $\binom{6}{2}\binom{6}{2}$

4.14(ii) $2\binom{6}{1}\binom{6}{2}$ **4.15(a)** $\binom{6}{4}\binom{7}{4}$

4.15(b) $\binom{2}{2}\binom{5}{2}\binom{6}{4} + \binom{2}{0}\binom{5}{4}\binom{6}{4}$, $225 \cdot 4 \cdot 3 \cdot 2$

5.1(a)(i) 60 **5.1(a)(ii)** 36

5.1(b) By FTA, express n as $n = p_1^{m_1} p_2^{m_2} \cdots p_k^{m_k}$. Then number of positive divisors is $\prod_{i=1}^{k}(m_i + 1)$.

5.2(i) $\binom{5}{2}\binom{7}{2}$ **5.2(ii)** $\binom{5}{2}\binom{6}{2}$ **5.2(iii)** $\binom{5}{2}\binom{4}{1}\binom{3}{1}$

5.2(iv) $\binom{12}{4} - \binom{5}{2}\binom{6}{2}$ **5.6** $\binom{7}{2}\binom{8}{2}$ **5.7** $\binom{8}{4}$

6.1 $\binom{6}{3}$ **6.2** $5! \cdot 6 \cdot 5 \cdot 4 \cdot 3$ **6.3** $2 \cdot \frac{10!}{3!}$

6.4 $\binom{12}{4}$ **7.1(i)** $\binom{55}{4}$ **7.1(ii)** $\binom{47}{4}$

7.1(iii) $\binom{55}{4} - \binom{46}{4}$ **7.1(iv)** $\binom{11}{1}\binom{43}{2}$ **7.1(v)** $\binom{27}{4}$, 0

7.2(i) $\binom{23}{7} - \binom{19}{7}$ **7.2(ii)** $\binom{19}{7}$ **7.3(i)** $\binom{7}{2}\binom{6}{2}$

7.3(ii) $\binom{5}{2}^3$ **7.4(i)** $\binom{33}{3}$ **7.4(ii)** $\binom{31}{3} - \binom{25}{3}$

7.4(iii) $\binom{36}{3}$ **7.5** $\binom{2006}{4}$

7.6 $\binom{16}{2} + \binom{11}{2} + \binom{6}{2}$ **7.7** $\binom{n+r-1}{r}$

7.8 $\binom{r-1}{n-1}$ **7.9** $\binom{14}{4}$ **7.10** $\binom{5}{2}\binom{8}{4}$

7.12 $4!$ **7.13(i)** $4!5!$ **7.13(ii)** $4!2$

7.13(iii) $4!2^5$ **7.13(iv)** $5!5!$ **7.14(i)** $7!$

7.14(ii) $3!4!7$ **7.15(i)** $4!2!$ **7.15(ii)** $4 \cdot 3 \cdot 4!$

8.1(i) 9^6 **8.1(ii)** P_6^9 **8.3(i)** 8^4

8.3(ii) 7^5 **8.4** $\binom{8}{4}\binom{4}{2}\binom{2}{2}3 + \binom{8}{2}\binom{6}{3}\binom{3}{3}3$

8.5(i) $n!$ **8.5(ii)** $\binom{n+1}{2} n!$

8.5(iii) $\binom{n+2}{3} n! + \frac{1}{2} \binom{n+2}{2}\binom{n}{2}n!$ **9.1** $\frac{10!}{5!2!}$

9.2 $\frac{8!}{3!2!}, \frac{5!}{2!}$ **9.3(i)** $\frac{15!}{4!5!6!}$ **9.3(ii)** $\binom{10}{4}\binom{11}{5}$

9.3(iii) $\binom{11}{5}\binom{9}{4}$ **9.4(i)** n^m **9.4(ii)** P_m^n

9.4(iii) $\binom{n}{m}$ **9.4(iv)** n^{m-1} **9.5(i)** $n!$

9.5(ii) $\binom{n+1}{2}n!$ **9.5(iii)** $\binom{n+2}{3}n! + \frac{1}{2}\binom{n+2}{2}\binom{n}{2}n!$

9.6 4^{10} **9.7** $n!, n(n+1)!, n(n+2)! + \binom{n}{2}(n+2)!$

10.2 25 **10.3** $k = 3; (1+x)^{14}; \binom{14}{4}, \binom{14}{5}, \binom{14}{6}$

12.1 $\binom{101}{6}$ **12.5** $\frac{(3n+1)!}{(n+1)(2n)!}$ **13.1** $2^{10} - 2 - \binom{10}{5}$

13.2(i) $\binom{10}{4}$ **13.2(ii)** P_3^{20} **13.2(iii)** $\binom{10}{2}\binom{8}{2}\binom{6}{2}$

13.3 5 **13.4** $10! - 2^9$ **13.5** 576

13.6(i) 16 **13.6(ii)** 40 **13.6(iii)** 336

13.7 813 **13.8** $2!3!6 \cdot 7 \cdot 8 \cdot 9$

13.9 $3!5 \cdot 6 \cdot 7 \cdot 8 \cdot 9$ **13.10** 499

13.11(i) 3^{15} **13.11(ii)** $\binom{15}{3}2^{12}$ **13.11(iii)** $\binom{15}{4}\binom{11}{5}$

13.11(iv) $2^{15} + 15 \cdot 2^{14} + \binom{15}{2}2^{13}$ **13.11(v)** $3^{15} - 3 \cdot 2^{14}$

13.11(vi) 120 **13.12(i)** 40 **13.12(ii)** 32

13.13 $2 \cdot 3^2 \cdot 5^4, 2 \cdot 3^4 \cdot 5^2, 2^2 \cdot 3 \cdot 5^4, 2^2 \cdot 3^4 \cdot 5, 2^4 \cdot 3^2 \cdot 5, 2^4 \cdot 3 \cdot 5^2$

13.14(i) $\binom{7}{3}$ **13.14(ii)** $\binom{7}{2}\binom{6}{1} + \binom{7}{2}\binom{6}{2} + \binom{6}{1}\binom{6}{3}$

13.15(i) $\binom{10}{2}$ **13.15(ii)** $\binom{10}{4}$ **13.16** $\frac{8!}{3!2!3!}$

13.17 15, 10 **13.18(i)** $3\binom{5}{2}$ **13.18(ii)** $\binom{15}{2} - 5\binom{3}{2}$

13.19 $4 \cdot 2^7 - 4$ **13.20(i)** $\binom{500}{2}$

13.20(ii) $\binom{166}{2} + \binom{166}{1}\binom{334}{1}$ **13.20(iii)** $\binom{166}{2} + \binom{166}{1}\binom{167}{1}$

13.21 $2^{n-1} - 1$ **13.22(i)** 3

13.22(ii) By FTA, express n as $n = p_1^{m_1} p_2^{m_2} \cdots p_k^{m_k}$. Then number of ways is $2^{k-1} - 1$. **13.24** $\binom{10}{6} 4^6$

13.25(i) 2^{10} **13.25(ii)** 2^5 **13.26(i)** $\binom{30}{4}$

13.26(ii) $\binom{33}{4}$ **13.27** 64 **13.28** $(n-1)!$

13.29(i) $\binom{15}{2} - \left(\binom{3}{2} + \binom{4}{2} + \binom{6}{2} + \binom{2}{2}\right)$

13.29(ii) $\binom{15}{3} - \left(\binom{3}{3} + \binom{4}{3} + \binom{6}{3}\right)$

13.29(iii) $\binom{15}{4} - \left(12\binom{3}{3} + 11\binom{4}{3} + 9\binom{6}{3} + \binom{4}{4} + \binom{6}{4}\right)$

13.29(iv) $\binom{15}{4} - \left(12\binom{3}{3} + 11\binom{4}{3} + 9\binom{6}{3} + \binom{4}{4} + \binom{6}{4}\right)$

13.30(i) $3^n - 2^n$ **13.30(ii)** $n(n-1)$ **13.30(iii)** $\binom{n}{3} 2^{n-3}$

13.31(i) $5 \cdot 4^3$ **13.31(ii)** $5 \cdot 4^3$ **13.31(iii)** $5 \cdot 4^2 \cdot 3$

13.31(iv) $5 \cdot 4 \cdot 13$ **13.31(v)** $5 \cdot 4 \cdot 3^2$ **13.31(vi)** $5 \cdot 4 \cdot 3 \cdot 2$

13.32(i) $8 \cdot 7 \cdot 6 + 7 \cdot 7 \cdot 6 \cdot 2$ **13.32(ii)** $7 \cdot 7 \cdot 6 \cdot 2$

13.32(iii) $8 \cdot 7 \cdot 6$ **13.32(iv)** $4 \cdot 8 \cdot 7 \cdot 6 + 2 \cdot 9 \cdot 8 \cdot 7 \cdot 6$

13.32(v) $9 \cdot 9 \cdot 8 \cdot 7 \cdot 6 - (4 \cdot 8 \cdot 7 \cdot 6 + 2 \cdot 9 \cdot 8 \cdot 7 \cdot 6)$

13.32(vi) $7 \cdot 8 \cdot 7 \cdot 6 + 9 \cdot 7 \cdot 6$

13.33(i) $\binom{12}{4}\binom{8}{1} + \binom{12}{1}\binom{11}{3}\binom{8}{1}$

13.33(ii) $\binom{12}{3}\binom{8}{1}\binom{7}{1} + \binom{12}{1}\binom{11}{2}\binom{8}{2}$

13.33(iii) $\binom{20}{1}\binom{19}{4} - \binom{12}{1}\binom{11}{4}$

13.33(iv) $\binom{12}{1}\binom{11}{4} + \binom{12}{4}\binom{8}{1} + \binom{12}{1}\binom{11}{3}\binom{8}{1} + \binom{12}{3}\binom{8}{1}\binom{7}{1} + \binom{12}{1}\binom{11}{2}\binom{8}{2}$

13.33(v) $\binom{19}{3}\binom{16}{1} + \binom{19}{4}$ **13.33(vi)** $\binom{19}{1}\binom{18}{4}$

13.33(vii) $2\binom{18}{3} + \binom{18}{1}\binom{17}{2}$ **13.33(viii)** $\binom{18}{1}\binom{17}{4}$

13.33(ix) $2\left(\binom{18}{3}\binom{15}{1} + \binom{18}{4}\right)$

13.33(x) $\binom{12}{4}\binom{8}{1} + \binom{12}{1}\binom{11}{3}\binom{8}{1} + \binom{12}{2}\binom{8}{2}\binom{6}{1} + \binom{12}{1}\binom{11}{1}\binom{8}{3} + \binom{12}{4}\binom{8}{1}$

13.34(i) $\binom{7}{3}$ **13.34(ii)** $\binom{7}{6}2^6$ **13.34(iii)** $\binom{7}{1}\binom{6}{4}2^4$

13.34(iv) $\binom{7}{2}\binom{5}{2}2^2$ **13.34(v)** $\binom{14}{6} - \binom{7}{6}2^6$ **13.35(i)** $\binom{6}{3}\binom{3}{2}$

13.35(ii) $\frac{\binom{6}{3}\binom{4}{2}}{3!}$ **13.35(iii)** $\frac{\binom{6}{2}\binom{4}{2}\binom{2}{1}}{2!2!}$ **13.35(iv)** $\binom{6}{2}\binom{4}{2}$

Index